Advances in

ECOLOGICAL RESEARCH

VOLUME 7

Advances in ECOLOGICAL RESEARCH

Edited by

J. B. CRAGG

Environmental Sciences Centre (Kananaskis), University of Calgary, Calgary, Alberta, Canada

VOLUME 7

1971

ACADEMIC PRESS
London and New York

ACADEMIC PRESS INC. (LONDON) LTD.
Berkeley Square House
Berkeley Square
London, W1X 6BA

U.S. Edition published by
ACADEMIC PRESS INC.
111 Fifth Avenue, New York 10003, New York

Library of Congress Catalog Card Number: 62–21479

ISBN: 0–12–013907–3

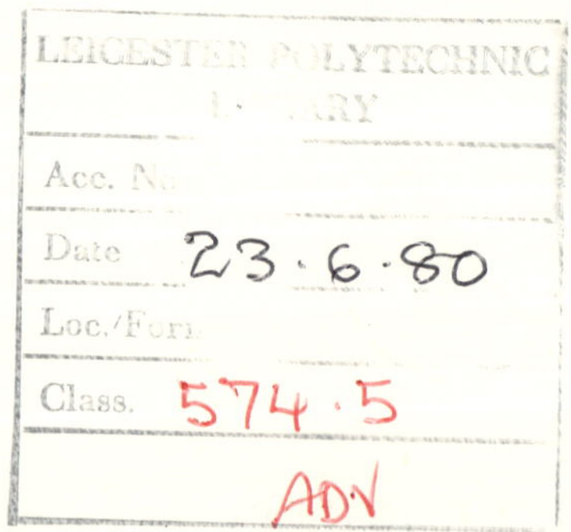

Printed in Great Britain by T. & A. Constable Ltd., Edinburgh

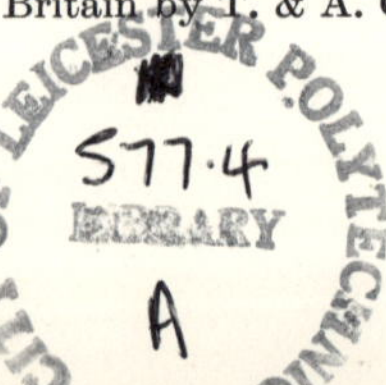

Contributors to Volume 7

J. Antonovics, *Department of Biology, University of Stirling, Scotland.*

C. C. Black, *Biochemistry Department, University of Georgia, Athens, Georgia, U.S.A.*

A. D. Bradshaw, *Department of Botany, University of Liverpool, Liverpool, England.*

J. R. Bray, *P.O. Box 494, Nelson, New Zealand.*

J. A. Gulland, *Fishery Economics and Institutions Division, Food and Agriculture Organization, Rome, Italy.*

R. G. Turner, *School of Plant Biology, University College of North Wales, Bangor, Caerns., Wales.*

Preface

The four reviews in this volume cover very different aspects of environmental biology as it relates to developments in ecology.

Dr Clanton C. Black's contribution is concerned with a matter of fundamental importance in ecology, the rate of net photosynthesis in plants. He draws attention to the differences in photosynthetic production capacity in different groups of plants and discusses the ecological implications of such differences.

The review of the tolerance to toxic heavy metals of certain plants by Messrs Antonovics, Bradshaw and Turner brings together information on the ability of plants and micro-organisms to combat such toxic effects. They consider both natural and man-made plant communities and the studies have considerable relevance in relation to present-day efforts to combat pollution.

Dr Gulland's paper on fisheries research is timely because results obtained from marine studies have often been ignored by terrestrial ecologists. Mathematical modelling has a long history in fisheries biology and its applications in the utilization of fish stocks have a bearing on general problems of conservation.

Dr Bray's study shows the effects of climatic trends on vegetational distribution, tree growth and crop success over the last two thousand years. His review summarizes physical, geophysical, glaciological and botanical data for both hemispheres and provides a palaeoecological basis for assessing some aspects of present-day patterns of vegetation.

April, 1971 J. B. Cragg

Contents

Ecological Aspects of Fishery Research

J. A. Gulland

Vegetational Distribution, Tree Growth and Crop Success in Relation to Recent Climatic Change

J. R. Bray

Heavy Metal Tolerance in Plants

J. ANTONOVICS

Department of Biology, University of Stirling, Stirling, Scotland, U.K.[1]

A. D. BRADSHAW

Department of Botany, University of Liverpool, Liverpool, England, U.K.

AND

R. G. TURNER

School of Plant Biology, University College of North Wales, Bangor,[2] *Caerns., U.K.*

[1] Present address: Department of Botany, Duke University, North Carolina, U.S.A.

[2] Present address: Shell Research Ltd., Woodstock Laboratories, Sittingbourne, Kent, U.K.

I. Introduction

The heavy metals, defined by Passow *et al.* (1961) as those metals having a density greater than five, include about thirty-eight elements. Their common feature in relation to biological life is that in excessive quantities they are poisonous and can cause death of most living organisms. However, certain organisms are remarkable in that they possess an ability to survive under conditions of metal contamination which would prove toxic to other living things. It is the purpose of this review to bring together the literature on the plants and micro-organisms which can combat excessive quantities of heavy metal ions.

Much of the literature on the subject is scattered, but in recent years a co-ordinated picture has started to emerge. The picture is by no means complete, and this review is as much an attempt to point to directions of investigation in the future as an attempt to collate the past work on this subject.

Toxic levels of heavy metals can occur under several circumstances.

Firstly, soil may itself contain large quantities of metal and it is with this type of contamination that the review will be mainly concerned.

The contamination results either from the presence of undisturbed metal ore near the soil surface causing "anomalies" as they are termed by the geochemist (e.g. Warren and Delavault, 1948; Duvigneaud and Denaeyer-de Smet, 1963; Tooms and Jay, 1964) or from the actual mining of ore bodies. Many waste products from mining activities are contaminated with metal at toxic levels and can produce large-scale pollution. Unproductive ore, tailings (see Bateman and Wells, 1917, and Griffith, 1919, for vivid descriptions) and seepage (e.g. Lackey, 1938) regularly contaminate mining areas. Smelting of ore produces contaminated slag and smoke fumes which may carry the contamination large distances (e.g. Ramaut, 1964; Hilton, 1967; Kerin, 1968). Other sources of soil contamination are less obvious and not so severe because they occur on a much smaller scale. For example, soil along main arterial roads frequently contains a high level of lead since this is an important component of "anti-knock" additive for petrol (e.g. Cannon, 1960a; Warren and Delavault, 1960, 1962; Suchodoller, 1967; Cholak *et al.*, 1968). The area below galvanized (zinc coated) fences and pylons may even have toxic zinc concentrations (Harris, 1946; Snaydon, quoted in Bradshaw *et al.*, 1965) and there may be accumulation of toxic amounts of lead on rifle ranges (McAllister, 1965). There is also evidence of increasing general environmental pollution resulting from widespread use of compounds of lead (Ruhling and Tyler, 1968, 1969) and mercury (Swedish Royal Commission, 1967).

Secondly, heavy metals are frequently used because of their toxicity as components of fungicides, pesticides or disinfectants. This source of contamination is important not only in that it may produce toxic levels of metals in the environment (Drouineau and Mazoyer, 1956; Gibson, 1958; Taschenberg *et al.*, 1961; Delas, 1963; Aomine *et al.*, 1967) and so affect the background vegetation and wild-life over which the spray is applied, but also in that the pathogen or pest over which control is being exercised may itself become resistant.

Thirdly, tolerant races can be produced in organisms growing in laboratory culture on artificial media containing heavy metals.

It is impossible to define precisely what is implied by the description "metal tolerant" because the phrase is generally used in two main senses. In one sense it refers generally to any species found occurring in an area of toxicity from which other species appear to be excluded. But it may be used more precisely to refer to specific individuals of a species which are able to withstand greater amounts of toxicity than their immediate relatives on normal soil. In the second case we are dealing with a precisely determined situation, a species normally non-tolerant but with an ability to evolve tolerant races. In the first case it will not be clear what is happening unless further investigations are made: either the species is already tolerant throughout its range (including uncontaminated sites) or it is an example of the second case and has evolved tolerant races. In those cases which have been examined it is always the latter that is found.

The theoretical problems raised by metal tolerant organisms fall into three main categories, which are separable only for the basis of argument.

1. What is the effect of heavy metal contamination on the flora (and fauna) in a given area? What organisms survive and what factors determine their distribution?

2. How has metal tolerance evolved? Is it an inherent property of some species, can it arise rapidly, and what processes are involved?

3. What is the mechanism of tolerance? How is it determined physiologically and biochemically?

The practical problems and implications of metal tolerance are several.

1. Prospecting for metal ore deposits can benefit from vegetation studies. Individual species or vegetation types may act as indicators of ore deposits. Some plants accumulate metals and deep rooted ones may bring up into their aerial parts metals which are not evident on the surface.

2. Micro-organisms play some part in the commercial extraction of metals from low-grade ores. Their role in contaminated substrates is therefore relevant.

3. Mining and smelting activities have created large areas of unsightly, and sometimes dangerous dereliction. The recolonization of these areas with vegetation is an important problem.

4. As a result of the wide use of heavy metals in sprays and other toxicants, the organisms that are being controlled may develop resistance which completely destroys the efficiency of the sprays. A knowledge of metal tolerance is necessary to minimize the chances of this happening.

The problem of heavy metal tolerance may appear to be a highly specialized topic, but it is hoped to show in this review that its study is not only of great interest and practical consequence, but also that it is a paradigm for many biological problems.

II. Ecology of Metal Tolerance

This section deals almost entirely with the plants found growing in the toxic soils that result from mining activity and natural "anomalies". The problem of races resistant to toxicants and laboratory strains will be dealt with in Section V (A2 and A3).

A. SPECIES PRESENT ON CONTAMINATED SOILS

No attempt will be made to produce comprehensive lists of species present on contaminated soils. The number of species mentioned in relation to such soils is immense and frequently there is little indication of whether they are actually growing on areas that are effectively toxic (see Section VIIA). A comprehensive list would therefore be of dubious value. This section summarizes the systems of classifying such plants, the main references to the various groups, and the most obvious features and species to emerge from the literature.

1. *Vascular plants*

Ferns are rarely mentioned in the literature relating to toxic areas but Vogt (1942b) mentions *Asplenium adulterium* as an indicator of nickel. Ferns are also known on serpentine soils which may be high in nickel and chromium (e.g. Kruckeberg, 1964). Wild (1968) records the ferns *Pellaea calomelanos* and *Chelianthes hirta* on copper and occasionally nickel soils in Rhodesia.

Similarly, there is no known record of Gymnosperms as consistent members of communities on contaminated soils, although they are used in biogeochemical prospecting; their leaves and twigs show enhanced metal content when growing over soil containing increased, but not toxic, levels of metal (e.g. Warren and Delavault, 1948).

Higher plants characteristic of metal-contaminated soils have been recognized for several centuries. Thalius, in 1588, noted *Minuartia verna* as a metal indicator (quoted by Ernst, 1965a) and even before 1900 there was considerable interest in these plants: for example, Williams (1830), Henwood (1857), Baumann (1885) and Jensch (1894) all quote species found consistently on metal-contaminated soils.

Plants that are largely restricted to or particularly abundant on metal-contaminated soils have been used as indicators of heavy metal ores and have therefore attracted special interest: their role in prospecting is more fully reviewed in Section VIIIA. Frequently the definition of a plant as an "indicator" is extremely vague, and often subjective. It is, generally speaking, a plant that in a given area or geographic region has been recognized as associated with a particular metal. Such plants are listed in Tables Ia and Ib.

Recently it has been realized by both the ecologists and biogeochemical prospectors that it is necessary to quantify the relative abundance of different species in relation to metal content (and other ecological factors) of the environment. The concept of indicator species is perhaps an ecologically unrealistic, although at times useful, qualitative assessment.

TABLE Ia

Plants that have been used as indicators in prospecting for heavy metals

Universal (U) or Local (L) Indicator	Species	Family	Metal	Locality	Reference
U	*Gypsophila patrini*	Caryophyllaceae	Cu	U.S.S.R.	Nestvetaylova, 1955
L	*Polycarpaea spirostylis*	Caryophyllaceae	Cu	Australia	Bailey, 1889
U	*Acrocephalus robertii*	Labiatae	Cu	Katanga	Duvigneaud, 1958
L	*Elshotzia haichowensis*	Labiatae	Cu	China	Tsung-Shan, 1957
U	*Becium* (= *Ocimum*) *homblei*	Labiatae	Cu	Rhodesia	Anon., 1959
U	*Merceya latifolia*	Moss	Cu	Sweden and Montana	Persson, 1948
L	*Eschscholtzia mexicana*	Papaveraceae	Cu	Arizona	Lovering *et al.*, 1950
L	*Armeria maritima*	Plumbaginaceae	Cu	Scotland	Henwood, 1857
U	*Tephrosia* sp. nov.	Leguminosae	Cu	Australia	Cole, 1965
L	*Polycarpaea glabra*	Caryophyllaceae	Cu	Australia	Cole, 1965
U	*Bulbostylis barbata*	Cyperaceae	Cu	Australia	Cole, 1965
U	*Fimbristylis* sp. nov.	Cyperaceae	Cu	Australia	Cole, 1965
L	*Loudetia simplex*	Gramineae	Cu	Rhodesia	Jacobsen, 1968
L	*Olax obtusifolia*	Olacaceae	Cu	Rhodesia	Jacobsen, 1968
L	*Erianthus giganteus*	Gramineae	Pb	Tennessee	Cannon, unpubl.
U	*Tephrosia* sp. nov.	Leguminosae	Pn, Zn	Australia	Cole, 1965
U	*Polycarpaea synandra* var. *gracilis*	Caryophyllaceae	Pb, Zn	Australia	Cole, 1965
L	*Tephrosia* affin. *polyzyga*	Leguminosae	Pb, Zn	Australia	Cole, 1965
L	*Gomphrena canescens*	Amaranthaceae	Pb, Zn	Australia	Cole, 1965
L	*Eriogonum ovalifolium*	Polygonaceae	Ag	Montana	Henwood, 1857
U	*Viola calaminaria*	Violaceae	Zn	Belgium and Germany	Vinogradov, 1955
L	*Philadelphus* sp.	Philadelphaceae	Zn	Washington	Yates, in Cannon, 1960a

An attempt to go beyond the simple indicator concept has been made directly in Belgium by Lambinon and Auquier (1964) who produced a semi-quantitative classification which was a considerable improvement on the crude concept of indicators. Their classification is summarized in Table II with some examples.

TABLE Ib

Plants that have been cited as indicators of metal containing soils but for which there is no clear cut evidence that they have been used for prospecting

Species	Family	Metal	Locality	Reference
Crotalaria cobalticola	Leguminosae	Co	Katanga	Duvigneaud, 1959
Silene cobalticola	Caryophyllaceae	Co	Katanga	Duvigneaud, 1959
Silene otites	Caryophyllaceae	Cu	Germany	Linstow, 1929
Viscaria alpina	Caryophyllaceae	Cu	Norway	Vogt, 1942a
Euphorbia matabelensis	Euphorbiaceae	Cu	Rhodesia	Jacobsen, 1967
Tapiphyllum velutinum	Rubiaceae	Cu	Rhodesia	Jacobsen, 1967
Combretum zeyheri	Combretaceae	Cu	Rhodesia	Jacobsen, 1967
Kyllinga alba	Cyperaceae	Cu	Rhodesia	Wild, 1968
Fimbristylis exilis	Cyperaceae	Cu	Rhodesia	Wild, 1968
Bulbostylis spp.	Cyperaceae	Cu	Rhodesia	Wild, 1968
Tephrosia longipes	Leguminosae	Cu	Rhodesia	Wild, 1968
Celosia trigyna	Amaranthaceae	Cu	Rhodesia	Wild, 1968
Vellozia equisetoides	Velloziaceae	Cu	Rhodesia	Wild, 1968
Hemizygia petrensis	Labiatae	Cu	Rhodesia	Wild, 1968
Andropogon amplectens	Gramineae	Cu	Rhodesia	Wild, 1968
Andropogon gayanus	Gramineae	Cu	Rhodesia	Wild, 1968
Astragalus declinatus	Leguminosae	Cu, Mo	Armenia, U.S.S.R.	Malyuga *et al.*, 1959
Alsine (= *Minuartia*) *verna*	Caryophyllaceae	Cu, Pb Ag, Zn	Europe	Linstow, 1929
Armeria vulgaris (cf. *halleri*, *elongata*, *maritima*, etc.).	Plumbaginaceae	Cu, Zn	Europe	Linstow, 1929
Amorpha canescens	Papilionaceae	Pb	Michigan, Wisconsin	Linstow, 1929
Rhus spp.	Anacardiaceae	Pb	Missouri	Linstow, 1929
Sassafras spp.	Lauraceae	Pb	Missouri	Linstow, 1929
Alsine setaceae	Caryophyllaceae	Hg	Spain	Linstow, 1929
Alyssum bertolonii	Cruciferae	Ni	Italy	Minguzzi *et al.*, 1948
Alyssum murale	Cruciferae	Ni	Georgia, U.S.S.R.	Doksopulo, 1961
Albizia amara	Leguminosae	Ni	Rhodesia	Wild, 1970
Dicoma macrocephala	Compositae	Ni	Rhodesia	Wild, 1970
Barleria aromatica	Acanthaceae	Ni	Rhodesia	Wild, 1970
Combretum molle	Combretaceae	Ni	Rhodesia	Wild, 1970
Dalbergia melanoxylon	Leguminosae	Ni	Rhodesia	Wild, 1970
Eminia antennulifera	Leguminosae	Ni	Rhodesia	Wild, 1970
Turraea nilotica	Meliaceae	Ni	Rhodesia	Wild, 1970
Pterocarpus rotundifolius	Leguminosae	Ni	Rhodesia	Wild, 1970
Lonicera confusa	Caprifoliaceae	Ag	Australia	Bailey, 1898
Trientalis europeae	Primulaceae	Sn	Bohemia	Linstow, 1929

TABLE Ib. contd.

Species	Family	Metal	Locality	Reference
Sempervivum soboliferum	Crassulaceae	Sn	Saxony	Linstow, 1929
Gnaphalium suaveoleus	Compositae	Sn	Brazil	Dorn, 1937
Thlaspi calaminare cf. (*alpestre*)	Cruciferae	Zn	Europe	Linstow, 1929
Silene vulgaris (= *inflata* = *cucubalus*)	Caryophyllaceae	Zn	Europe	Linstow, 1929
Arabis halleri	Cruciferae	Zn	Germany	Linstow, 1929
Thlaspi cepeaefolium	Cruciferae	Zn	Austria, Italy	Linstow, 1929
Anagallis collina	Primulaceae	Zn	Italy	Linstow, 1929
Cistus monspeliensis	Leguminosae	Zn	Sardinia	Linstow, 1929
Ruta graveoleus	Rutaceae	Zn	Brazil, U.S.A.	Dorn, 1937
Ruta latifolia	Rutaceae	Zn	Brazil	Dorn, 1937
Senecio brasiliensis	Compositae	Zn	Brazil	Dorn, 1937
Matricaria americana	Compositae	Zn	Brazil	Dorn, 1937
Populus deltoides	Salicaceae	Zn	U.S.A.	Buck, 1949
Ambrosia spp.	Compositae	Zn	U.S.A.	Buck, 1949

Plants regarded as indicators would either be metallophytes or extreme examples of elective pseudometallophytes. Malyuga (1964) has distinguished two types of indicator, namely "universal" which corresponds to the category absolute metallophyte, and "local" which corresponds to the category local metallophyte.

Recently Jacobsen (1968) from experience with a wide range of species in Rhodesia, as a result of a need to follow the extent of a copper-bearing reef, has attempted to define indicator species more precisely. He has determined the highest and lowest levels of soil copper at which individual species are found and from this calculated the specific indicator value for each species from

$$\frac{\text{highest Cu level} - \text{lowest Cu level}}{\text{average Cu level}}.$$

From this he considers good indicators to be species with an indicator value of 4 or less arising either from occurrence over a small range of copper levels or from occurrence at high average copper levels. Typical species include *Bulbostylis contexta* and *Eragrostis racemosa*. Although somewhat imprecise as used at present it is a technique that indicates very clearly how different species may be distributed in relation to different levels of metal in the soil.

TABLE II

Classification of plants on metal contaminated soils (after Lambinon and Auquier, 1964) with examples from metal contaminated soils of Europe

Category	Examples
1. *Metallophytes*—taxa found only on metal contaminated soils.	
(a) *Absolute metallophytes*— found only on metal contaminated soil over all their distribution.	e.g. *Viola calaminaria*, *Thlaspi alpestre* ssp. *calaminare*, *Minuartia verna* ssp. *hercynica*.
(b) *Local metallophytes*— found only on metal contaminated soil within a given region but occurring also in a phyto-geographically distinct non-contaminated area.	e.g. *Armeria maritima*.
2. *Pseudometallophytes*—taxa occurring both on contaminated soils and on normal soils in the same region.	
(a) *Elective pseudometallophytes*— abundant and often more vigorous on contaminated soil.	e.g. *Agrostis tenuis*, *Campanula rotundifolia*, *Polygala vulgaris*, *Thymus pulegioides*, *Rumex acetosa*.
(b) *Indifferent pseudometallophytes*— live on contaminated soil regularly but show neither abundance nor particular vitality.	e.g. *Plantago lanceolata*, *Avena pubescens*, *Genista tinctoria*, *Linum catharticum*.
(c) *Accidental metallophytes*— usually weeds and ruderals appearing sporadically and showing reduced vigour on metal contaminated soils	

In the classification of Lambinon and Auquier (1964) the category in which any particular species is placed depends on the locality or group of localities being investigated, since the relative abundance of species on and off contaminated soils depends on other factors in addition to metal concentration (see Section IIc). This is elegantly illustrated by recent work on *Becium homblei* (Howard-Williams, 1970). This is an excellent indicator of copper in Southern Rhodesia, and has been considered an absolute metallophyte. But Howard-Williams (1970) shows it to be a local metallophyte, widely distributed on non-copper soils in Northern Rhodesia and not always restricted to copper soils in Southern Rhodesia.

The classification described by Lambinon and Auquier (1964) has been used by Auquier (1964) and by Demoulin *et al.* (1967). A related classification is used by Duvigneaud and Denaeyer-de Smet (1963), who describe in detail the vegetation on naturally occurring copper outcrops in Katanga.

Phytosociological studies in Germany (Braun-Blanquet, 1951; see also Poore, 1955a, b, c, 1956) have provided what is probably the best semi-quantitative system of classifying plant communities, and the techniques have been applied to communities on contaminated soils (see Section IIB). The method of classification is based on vegetation composition of the communities and hence provides an indirect means of classifying species characteristic of metal-contaminated soils.

It is appropriate here to give a brief outline of the phytosociological technique of classifying communities, since this has been an important technique both in providing lists of species characteristic of metal-contaminated soils and in classifying communities carried on such soils (Section IIB).

In these techniques the basic unit of classification is the "association". The associations can either be further subdivided (into e.g. sub-associations and varieties) or they can be grouped into families, orders, and classes according to their affinities. The process whereby associations are recognized initially involves describing by means of "releves" (species lists and notes on environmental factors) a series of "stands" (areas of vegetation to be classified). These "releves" are then sorted into groupings by means of the presence of sets of species which one group has in common ("differential" species or "Trennart"). If a group of releves is considered sufficiently distinct from other associations or other groups of releves, then the particular group is given association rank.

The differential species may be found in other unrelated associations, or they may be entirely restricted to that association. The latter group of species are known as "characteristic" species or "Kennart" and are used to recognize the communities: they are the indicators of particular associations.

Confounded with the concepts of differential and characteristic species are the concepts of "constancy" and "fidelity". A species has a high constancy if it occurs in most of the releves of an association (e.g. constancy is V if it occurs in 81–100% of the releves). "Constancy" therefore refers to the frequency of occurrence of species within an association. A differential species is one with a high constancy in a particular association, but low constancy in others. "Fidelity" refers on the other hand to the distribution of species among associations. There are five classes of fidelity: (V) exclusive species confined almost completely to one community; (IV) selective species found most frequently in certain communities but also rarely in others; (III) preferential species present in several communities more or less abundantly but predominantly in one and there with greater vigour; (II) indifferent species without any particular affinity for any particular community; (I) accidental species which are rare intruders or relics from another community. This latter classification is clearly related to that of Lambinon and Auquier (1964).

The Braun-Blanquet method therefore provides us with a means of indicating the relative abundance of different species in communities on metal contaminated soil (constancy and differential species) and their relative restriction or otherwise to such soils (fidelity, characteristic species and differential species).

These techniques have been used by several workers in Germany to classify the communities on contaminated soils (see Section IIB for full references). In so doing these workers have provided extensive species lists of plants on contaminated soils. It is not possible here to list all the species given by these workers, since they are too numerous and often it is not clear to what extent the soils on which some species are growing are actually toxic. However, Ernst (1965a, 1968a) has summarized the species on these soils in Europe which are characteristic of or differential to the communities (see Table III). Certain species are clearly unique components of most metal-contaminated soils, others are restricted to particular metal-contaminated communities, whereas a third group are consistently present in some communities but by no means confined to metal-contaminated soils.

All the methods of classifying species occurring on metal-contaminated soils are compared in Table IV.

2. *Mosses and Liverworts*

Several species of mosses (often belonging to the genera *Merceya* and *Mieliochhoferia*) have been given the name of "copper mosses" (Persson, 1948, 1956; Noguchi, 1956; Noguchi and Furuta, 1956): they have been used in prospecting (Persson, 1948) and it has been postulated, though with no evidence, that they are restricted to copper soils by virtue of their being able to use hydrogen from hydrogen sulphide into photosynthesis (Schatz, 1955).

Warncke (1968) observed that *Marchantia alpestris* was generally restricted to mineral-, especially copper-rich areas in Scandinavia: in Denmark it was confined to an area contaminated with copper and zinc from effluent of a sulphuric acid plant. He suggested that it too should be included under the category of "copper mosses". Further studies of mosses on mineral-rich soils have been made by Schacklette (1965), who showed that they may sometimes accumulate heavy metal ions.

Numerous species of mosses are recorded as members of communities on metal-contaminated soils (Schwickerath, 1931; Schubert, 1954a, b;

TABLE III

Differential and characteristic species of West and Central European plant communities of heavy metal soils, showing their constancy in various associations (after Ernst, 1966)

Family:	Armerion		Thlaspeion		Galio-Minuartion	
Association:	*Armerietum bottendorfensis)* (Schubert, 1952)	*Armerietum halleri* (Libbert, 1930) in Ernst, 1966	*Violetum calaminariae* (Schwickerath, 1931)	*Armerietum muelleri* ass. nov.	*Violetum dubyanae* (Ernst, 1965a)	*Thlaspeetum cepeaefolii* (Ernst, 1965a)
Number of stands:	29	78	41	14	15	10
Armeria maritima ssp. *bottendorfensis*	V	...	...	...	...	...
Thymus chamaedrys	V	...	...	...	...	...
Poa badensis	IV	...	...	...	...	...
Festuca glauca	IV	...	...	...	...	...
Festuca sulcata	IV	...	...	...	...	...
Armeria maritima ssp. *halleri*	...	V	...	...	...	...
Cladonia alcicornis	V	III	...	...	...	...
Asperula cynanchica	V	III	...	...	...	...
Silene otites	IV	II	...	...	...	...
Scabiosa canescens	III	III	...	...	...	...
Potentilla heptaphylla	III	II	...	...	...	...
Viola calaminaria	...	...	V	...	...	...
Armeria maritima ssp. *calaminaria*	...	...	III	...	...	...
Armeria maritima ssp. *muelleri*	...	...	...	V	...	...
Thlaspi alpestre ssp. *calaminare*	...	...	IV	V	...	...
Festuca ovina ssp. *ovina*	I	IV	IV	V	...	...
Agrostis tenuis	III	V	III	V	...	...
Viola dubyana	...	...	...	...	V	...
Euphrasia salisburgensis	...	...	...	...	V	...
Thymus alpigenus	...	...	...	...	V	...
Thlaspi cepeaefolium	...	...	...	...	...	V
Galium anisophyllum	...	...	...	...	IV	IV
Poa alpina	...	...	...	...	IV	V
Dianthus silvester	...	...	...	...	I	III
Silene cucubalus var. *humilis*	I	V	IV	V	IV	V
Minuartia verna ssp. *hercynica*	V	IV	III	V	V	IV

Constancy classes 0–20% = I, 21–40% = II, etc., refer to number of stands showing that species in the association.

Characteristic species (underlined) = species restricted to that association.

Differential species (not underlined) = species generally common to the stands in an association.

TABLE IV

Comparison of various classifications of plants on metal-contaminated soils

Malyuga (1964)	General ecological terms	Lambinon and Auquier (1964)	Duvigneaud and Denaeyer-de Smet (1963) and Wild (1968)	Braun-Blanquet		
				Fidelity	"Type" species	Constancy
Indicator	Endemic or Bodenstet	Metallophyte	Metallophyte	Exclusive	Characteristic	Refers to the consistency with which any of the previous categories occur in a community. Characteristic and differential species have generally a high constancy.
Universal		Absolute	Eumetallophyte	Selective	Differential	
Local		Local	Local metallophyte			
	Bodenwaag	Pseudometallophyte				
		Elective Indifferent	} Metallophile	Preferential Indifferent		
		Accidental		Accidental		
			Metalloresistant*			

* A separate category of ubiquitous species defined by the fact that they do not take up metal.

Lambinon and Auquier, 1964; Ernst, 1965a, 1968b), and Url (1956) studied the copper resistance of several species from a copper-contaminated region in Austria. Among the more commonly mentioned types are *Bryum caespiticum*, *Pohilia nutans* and spp., and *Weisia* spp.

Liverworts are far rarer on contaminated soils. *Cephaloziella* spp. are recorded by Koch (1932), Lambinon and Auquier (1964) and Ernst (1965a). The only other records are sporadic: there is mention of *Plectocoela crenulata* (McAllister, 1965) and *Riccia bischoffii* (Schubert, 1954b).

3. *Lichens*

Lichens are found both as colonizers of bare rock and as general components of the established vegetation on contaminated soils. Poelt (1955), Lampe and Klement (1958) and Lange and Ziegler (1963) consider the various encrusting lichens found as initial colonizers on bare rock associated with mining debris. The genera *Acarospora* and *Lecanora* are frequently mentioned; Lange and Ziegler (1963) show these and other species to contain large quantities of iron and copper mainly in the cell walls. Wild (1968) notes that a species of *Lecanora* is found on natural rocky outcrops in Rhodesia with its thallus actually growing over exposed malachite without any apparent harmful effects. This is remarkable if we consider that the algal component at least might be expected to be particularly sensitive to copper toxicity.

Other lichens are listed in work on community studies on contaminated soils. Ernst (1965a) quotes three *Cladonia* species and *Cornicularia aculeata* as characteristic species of sub-associations. *Cladonia* spp. and *Stereocaulon* spp. (Schwickerath, 1931; Lambinon, 1964; Lambinon and Auquier, 1964; McAllister, 1965; Ernst, 1965a, 1968a) appear as consistent components of mine communities. Maquinay *et al.* (1961) show that *Stereocaulon nanodes* Tuck f. *tyroliense* (Nyl) M. Lamb may accumulate high levels of zinc (3300 ppm dry weight on soils with only 700 ppm).

4. *Micro-organisms, algae and fungi*

The general occurrence of micro-organisms in metal-contaminated soils is poorly documented although some of the best examples of bacteria which resist extreme conditions (for review see Brock, 1969) are those found in acid mine drainage. The most common species which are able to resist extremes of pH and copper and iron concentration are *Thiobacillus ferro-oxidans*, *Th. thio-oxidans* and *Ferrobacillus ferro-oxidans* (Bryner *et al.*, 1954). These bacteria, especially *Th. ferro-oxidans*, have been subsequently shown to be commercially important in leaching copper from low-grade sulphide ores (see Section VIIIB). The ability of these bacteria to resist extreme conditions is illustrated by the work of Booth and Mercer (1963) who showed that *Thiobacillus thio-oxidans* and *Ferrobacillus ferro-oxidans* could withstand 20 000 and 10 000 ppm of copper ions respectively. Ehrlich (1963b) reported the presence of not only bacteria of the *Thiobacillus–Ferrobacillus* group in mine water (pH 2·5, copper concentration 0·8 g/litre and iron concentration 1·06 g/litre), but also other organisms including yeasts (resembling *Rhodotorula* and *Trichosporon*) and Protozoa (an amoeba species and a flagellate resembling *Eutrepia*). It is clear that a wide variety of organisms inhabit mine water, and that a dominant role is played by chemotrophic bacteria in providing an initial carbohydrate source. As Ehrlich (1963b) states, "The microbial population in the mine water sample suggests the existence of a balanced ecological system in which carbon fixation is dependent on chemosynthetic and not photosynthetic autotrophy".

Subsequent work has shown fungi to be consistent if not abundant components of mine soils. The ability of fungi to withstand extreme conditions was illustrated by Starkey and Waksman (1943), who showed that *Acontium velatum* and another unidentified species could grow in pH 0 and 4% $CuSO_4$ solution. Several workers have isolated micro-organisms tolerant to high metal levels from contaminated soils. Seal (1970) showed that fungi (*Aspergillus* spp.) from copper-mine soil have a greater copper tolerance than related fungi from normal soil. The work reported in Hilton (1967, p. 92) demonstrated that the microbial population of contaminated slag tips was very sparse, but increased greatly with the addition of organic matter. Tonomura, Nakagami, Futai and Maeda (1968) isolated mercury resistant bacteria from soils contaminated with mercury-containing fungicides. Williams (1970) has isolated at least six different Actinomycetes from very acid copper-mine waste, and noted the occurrence of fungi and bacteria. And Demoulin *et al.* (1967) record the occurrence of three Gasteromycetes *Scleroderma bovista* Fr., *Lycoperdon spadiceum* Pers. and *Bovista plumbea* Pers. on calamine soils often containing well over 5000 ppm zinc.

Records of algae on heavy metals are rare but Hassall (1962) notes that *Chlorella vulgaris*, a green alga, is tolerant to barium, manganese, lead and copper. Duvigneaud (1958) records an algal substratum of blue-green algae dominated by *Cirosiphon geniculatus* in cupriferous soils in Katanga (Congo). Such colonies have also been noted by Wild (1968) in Rhodesia where some acute and experienced prospectors have actually used them as copper indicators since they form characteristic black crusts at the soil surface. In this case, species of blue-greens belonging to the genera *Scytonema* and *Microcoleus* seem to be dominant but other blue-greens such as *Nostoc*, *Phormidium*, *Anabaena* and *Oscillatoria* also occur, as well as a diatom (*Hanzscha* sp.), and a green alga belonging to the genus *Bracteacoccus*. In Britain similar algal crusts are readily observable on copper and zinc mine soils, and pools heavily contaminated with metals can have vigorous growth of filamentous green algae. In marine algae *Ectocarpus* (Russell and Morris, 1970) and *Enteromorpha* (Hodgson, 1969) can be found growing on ships treated with antifouling paints (see Section VA2).

High concentrations of heavy metals do not, therefore, prevent the growth of micro-organisms in contaminated soils: indeed some micro-organisms appear to be able to withstand metal concentrations which are toxic to even the most tolerant higher plants. However, the general nature of the microbial communities in metal-contaminated soils is poorly understood. Such communities are of interest because they appear often to rely on chemotrophic bacteria for their energy source and because their presence appears to promote release of metal ions. This is clearly important to the ecology of organisms growing on such soils.

5. *Conclusion*

The species found on toxic soils are very varied and differ according to the local ecological conditions and geographical area. No clear-cut taxonomic pattern emerges (see Section VB).

There appears to be a close affinity between the flora of toxic soils and serpentine soils (Rune, 1953; Spence, 1970). Serpentine soils are often high in nickel and chromium, and these metals appear in part at least to determine the serpentine flora (Birrell and Wright, 1945).

However, Walker (1954) working on serpentine soils in California considered that there was insufficient evidence that nickel or chromium toxicity was of importance as compared with high magnesium/calcium levels. Soane and Saunder (1959) on the contrary produce evidence that the Great Dyke serpentine soils of Rhodesia induce intense symptoms of nickel toxicity in oats. Furthermore, the importance of the effect of nickel, in some serpentine soils, on natural vegetation, is brought out by Wild (1970). After showing that the serpentine soils of the Great Dyke produce a depauperated and stunted flora he points out that if nickel anomalies occur in serpentine mother rocks, they are characterized by an even more depauperated flora derived from the already anomalous serpentine vegetation. This makes the nickel flora under these circumstances difficult to distinguish from that of the surrounding serpentine soils without high nickel values, but this can nevertheless be done by observing rather subtle differences in the proportions and relative dominance of certain species and the presence of a few indicator species either absent or rare on normal serpentine soils. Examples are *Albizia amara*, *Barleria aromatica*, *Combretum molle*, and an undescribed species of *Dicoma* (see Section IV) which is almost exclusively endemic on Rhodesian nickel soils.

Most of the higher plants found on toxic soils are perennial herbs. The perennial habit probably makes colonization easier since it ensures persistence and a low growing habit is an adaptation to exposure.

The study of species found on contaminated soils is difficult: often it is not possible to delimit precisely the contaminated area, to determine the degree of effective toxicity, or to assess the other ecological factors involved. Studies of species on metal-contaminated soils have rarely been rigorous and are usually little more than incomplete species lists. It is often possible that the species listed are not growing on metal-contaminated soils but on distinctive, poor soils adjoining contaminated areas. From the available information it is difficult to deduce generalities about taxonomic, morphological or physiological characteristics of the species involved since they obviously occupy a wide and diverse spectrum of ecological conditions.

It is none the less clear that the plants growing on metal-contaminated soils are often characteristic of such soils. Toxic areas are usually colonized by some plants, and although the colonization of many areas is sporadic, other areas carry regular communities.

B. COMMUNITY STUDIES ON CONTAMINATED SOILS

Several workers have examined the plant communities found on metal-contaminated soil using the techniques of phytosociology pioneered by Braun-Blanquet (see Section IIA1 for full explanation). These techniques enable plant communities to be classified in a way analogous to the classification of species. The species of phytosociology is the association, and associations can be grouped into families, orders and classes.

Schwickerath (1931), Koch (1932), Schubert (1953, 1954a), Ernst (1965a, 1968a) and Baumeister (1967) have examined the communities growing on contaminated soils (zinc and copper) in Germany. This work has been extended to Holland by Heimans (1936), to France by Ernst (1966), and to Great Britain by Shimwell (1967) and Ernst (1968b, c). These workers have shown that plants on contaminated soils form definite and distinct associations, which can be easily identified and named. Keys identifying communities of a particular country or area frequently include mention of associations on metal-contaminated soils (e.g. Tuxen, 1937; Lebrun *et al.*, 1949). A full bibliography is given by Ernst (1967).

Schwickerath (1931) was the first to study in detail communities on toxic soils (lead/zinc mines near Aachen, Germany). He named these as belonging to the association *Violetum calaminariae* (order *Mesobrometum*, class *Festuco-Brometa*). He further subdivided these into vegetation types or "facies", which he regarded as representing different ecological conditions. Koch (1932) in a brief study of communities on zinc-contaminated soil near Osnabruck, Germany, considered these to belong to another association, the *Minuartia verna–Thlaspi alpestre* association. Schubert (1953, 1954a, b), studying communities on lead, zinc and copper soil, recognized further associations, *Armerietum bottendorfensis* and *Armerietum halleri*, characterized by species of the genus *Armeria*.

Heimans (1936) noted the similarity of the vegetation in Holland with the *Violetum calaminariae* of Schwickerath (1931) with the exception that *Minuartia verna* was absent, and *Agrostis vulgaris* (= *tenuis*) a dominant member.

Ernst (1965a) reconsidered the classification of mine-plant communities by previous workers. All these associations had several species in common, namely *Viola calaminaria*, *Thlaspi alpestre*, *Minuartia verna*, *Silene vulgaris*, *Armeria* spp. and *Festuca ovina*, but occasionally one or other was absent. He came to the conclusion that the similarities between these communities was so great and they were so unlike any other communities, that they should all belong to one separate class (containing only one order), the *Violetea calaminariae*. This class and order could be identified by the consistent members *Silene vulgaris* and *Minuartia verna*. Ernst (1965a), as a result of additional observations, further divided this class into three families.

1. The *Armerion halleri*, of drier habitats in central Germany.
2. The *Thlaspeion calaminariae*, of western central Europe.
3. The *Galio-Minuartion vernae*, of the alpine regions.

Ernst (1966) recognized another association in southern France, but this could be included into the *Thlaspeion calaminariae* group. The mine vegetation of Britain (Ernst, 1968b, c) could be placed within the same family, although the communities in certain areas in particular were generally species poor. A summary of the characteristic species of these associations and families is given in Table III.

The work of Ernst demonstrates clearly the power of the phytosociological methods in summarizing the variation among communities on a wide range of contaminated soil types. Certain consistent patterns emerge, related in part to habitat and geography but the reasons for the existence of such regular but different communities remain unexplained.

C. FACTORS DETERMINING PLANT DISTRIBUTION ON CONTAMINATED SOILS

1. Metal concentration and type

The overriding characteristic of contaminated soils is the high concentration of metals in the soil. However, although the area of contamination is frequently recognizable by a flora different from that in surrounding areas, very few workers have studied precisely to what extent metal concentration and type are important in determining the distribution of mine plants.

Part of the difficulty arises from the problem of deciding on what measure to use for the amount of metal in the soil, since the metal may be in several forms. It may be water soluble, exchangeable and bound to inorganic or organic soil components, or unavailable in stable complexes with organic substrates, or unavailable as insoluble compounds or minerals (e.g. Nicolls *et al.*, 1965).

Most workers have resorted to measuring total metal concentration since broadly speaking this has been found to reflect in a relative way on the concentration affecting the plants, and because it is easy to measure (Duvigneaud and Denaeyer-de Smet, 1963; Gregory and Bradshaw, 1965; Nicolls *et al.*, 1965).

There is no evidence that other methods (e.g. water soluble, acid soluble) reflect more accurately the amount available to plants, and at present total metal content appears the simplest and most realistic measure. The use of bioassay to solve this difficult problem has not been explored: it may be possible to use growth of genotypes of, say, *Agrostis tenuis* of differing degrees of tolerance on soils to estimate the effective concentration at particular sites in a region.

Another complicating factor is the vertical distribution of metals in the soil. This is frequently non-uniform because of leaching, accumulation of metals in humus layers, localization of ore bodies at certain depths, or spread of toxic debris on normal soils (see Malyuga, 1964, for full discussion). Clearly plants with differing root systems may encounter effectively different concentrations of toxic metals.

Soils with very high concentrations are usually barren whereas those with only slightly enhanced concentrations may not visibly affect the vegetation but may lead to increased metal content of the plants growing there (see Section VIIIA). Different plants can withstand (in the sense of tolerate or evolve tolerance of) different amounts of metal. Schwickerath (1931) recognized several vegetation types representing increasing colonization, and these could be correlated with both lead and zinc concentration of the soil. Duvigneaud and Denaeyer-de Smet (1963) and Wild (1968) recognized three types of "cuprophyte" (plants restricted to copper soils) on natural and man-made copper soils of Katanga and Rhodesia respectively. They were the "polycuprophytes" which grew in areas of 5000–10 000 ppm copper or above, the "oligocuprophytes" which were restricted to soils with 800–2000 ppm copper, and the "eurycuprophytes" which could colonize a whole range of copper concentrations. The latter were relatively rare. Ernst (1965b) recognized three initial phases in the colonization of zinc-contaminated areas: and the last of these phases which contained a considerable number of species on soils with a generally lower concentration than the first two.

The study of Nicolls *et al.* (1965) emphasizes, in a subtle way, the controlling effect of metals on vegetation. As expected, sharp changes in the vegetation coincided with changes in total contamination. However, changes in the relative amounts of lead, copper and zinc were also important. Of the various species considered, *Eriachne mucronata* seemed to tolerate high concentrations of all three metals, *Bulbostylis barbata* and *Polycarpea glabra* were found in areas of high copper, while *Tephrosia* sp. nov. was found where the copper concentration was lower. In general, changes in lead and copper were more determinant than changes in zinc.

2. *Other soil factors*

Soils contaminated with heavy metals differ from each other and from normal soils in many ways other than in metal content. The nutrient status of the soil, the organic matter content and the texture may all affect the number and types of plant growing on the soil. Their effect may be direct or they may act indirectly by influencing the toxicity of the metal in the soil.

In general, mine-tip soils are extremely low in the essential nutrients nitrogen, phosphorus and potassium. Bradshaw *et al.* (1960), and McNeilly (1965) have shown that lead- and copper-tolerant populations of *Agrostis tenuis* are also better able to grow in culture solutions containing low phosphate than are populations from pasture soils. Recent studies on reclamation of contaminated tips (Hilton, 1967; Smith and Bradshaw, 1970) (see also Section VIIIc) have shown that addition of complete fertilizer at normal agricultural rates enormously improves the growth of plants both native and foreign to mine soils: in many cases, native tolerant plants can be made to grow on previously bare mine soils only with the addition of complete fertilizer. Further indirect evidence comes from distribution studies. Horne (1967) found that greatest vegetation cover on a small area of a copper mine could best be correlated with high potassium content. Nicolls *et al.* (1965) state that phosphorus may determine the distribution of plants on a natural metal (copper and zinc) outcrop in Australia: in contrast to mine tips, the soil there had a higher phosphorus content than that of surrounding areas, and comparable nutrient content in potassium and nitrogen.

Mine-tip habitats are rich in sulphur since the metals usually occur as sulphur ores. It has been shown (Antonovics *et al.*, 1967; Turner, 1967) that *Agrostis tenuis* from a copper mine requires more sulphur for normal growth than the same species from pasture, and that sulphur oxidizing bacteria can be found in such soils (see Section IIA4).

Another element which appears to be important is calcium, either by virtue of its direct effect, or by its influence on soil pH. In Great Britain and Europe mines on acid soils differ markedly from those with a higher pH. They have generally fewer species and are usually covered with *Agrostis tenuis* and *Festuca ovina*; calcareous mines in contrast have *Agrostis stolonifera* and *Festuca rubra*. Liming of waste tips improves the performance of plants sown on mines, if fertilizer or organic matter are also added (Hilton, 1967). Schwickerath (1931) observed that his vegetation types representing different degrees of colonization, although they correlated with total lead and zinc in the soil seemed to correlate better with the calcium: zinc ratio. Since calcium may alleviate the uptake and toxic effects of zinc, this ratio is possibly a measure of zinc availability.

The importance of calcium has since been observed by other workers. Warren and Delavault (1949) found the zinc and copper content of trees growing over limestone areas was much lower than in more acid soils and concluded that "it is reasonable to assume that pH of soil is a more important factor than either the zinc or copper content of the ground water in determining the amount of zinc and copper which may be toxic". Nicolls *et al.* (1965) note that over calcareous rocks of a naturally occurring outcrop of mineralization of zinc and copper, there are different and more numerous herb species. Plants such as *Suaevola densevestita* and *Ptilotus obovatus* are listed as possible indicators of copper anomalies in calcareous areas.

The organic matter content of the soil is also of extreme importance since it forms stable complexes with metal ions, so making them unavailable (Hodgson *et al.*, 1966). Dykeman and de Sousa (1966) record concentrations of 7% copper in peat of a copper swamp yet it appears to have no toxic effects, the overlying vegetation being typical for the area. Organic matter (sewage sludge or domestic refuse) is extremely effective at alleviating the toxicity of metals in slag heaps, permitting healthy growth of various non-tolerant species for several years (Hilton, 1967).

The importance of yet other soil factors was pointed out by Ernst (1965b). He studied the ecological conditions on mines characterized by three initial stages of colonization, namely the *Silene cucubalus* (=*vulgaria*) var. *humilis* stage, the *Minuartia verna* ssp. *hercynicium* stage and the *Euphrasia* spp. stage. *Silene* grew on soils with a coarse texture, low water capacity, and low surface temperature. *Minuartia* on the other hand grew on soils with finer texture, high water capacity, but high surface temperature. The habits of these plants, namely the deep rooting *Silene* with narrower leaves than the normal forms, and the shallow-rooting but tufted habit and needle-like leaves of *Minuartia*, adapt them to their respective habitats. The *Euphrasia* phase was considered to be a succession from the above two phases, but also characteristic of soils of even higher water capacity and, as mentioned already, lower metal content. Ernst therefore provides very clear-cut evidence for the importance of factors other than metal content to plants growing on metal-contaminated soils, although the specific factor, water or nutrients, is not identifiable.

3. Climatic factors

Contaminated areas are frequently only sparsely colonized. The plants on such areas are therefore growing exposed in open communities. Several workers (see Section VF) have noticed morphological differences between plants growing on mine soils and those in their normal habitat. Usually these morphological differences are in the direction of plants on the mine being smaller, prostrate and apparently adapted to exposed conditions. Occasionally spreading forms are found, adapted to the sand-dune-like conditions of some tips made of tailings (Bradshaw, 1959). Bröker (1963), and Antonovics and Bradshaw (1970) have shown that these morphological differences are not pleiotropic effects of the tolerance mechanism but that they must have arisen by independent selection. This may well be selection for adaptation to climatic factors but it is possible in some instances that the small size is an adaptation not so much to exposure but to the nutrient shortage in the habitat.

As would be expected, climatic factors exert considerable control on the species to be found in different contaminated areas. Wild (1968) shows that the species found on copper soils in the neighbouring areas of Katanga and Rhodesia are different because of the greater aridity of Rhodesia: for instance, *Combretum* spp. are common in copper soils in Rhodesia and almost absent in Katanga. For this reason, Malyuga (1964) divides Russia into nineteen biogeochemical zones.

4. Biotic factors

Plants probably play an important part in changing the nature of the toxic habitat. Their root systems stabilize the soil and their aerial parts offer some shelter to subsequent colonizers, but above all they contribute organic matter and humus to the surface layers of the soil. The addition of humus has three results: it increases the nutrient status of the soil, it forms complexes with heavy metals and makes metal ions unavailable to plants, and it improves soil texture. The ability of organic matter to complex with heavy metals frequently results in an accumulation of heavy metals at the surface of the soil and in humus layers of the soil (Dykeman and De Sousa, 1966; Le Riche, 1968). The cyclic process of mineral absorption, death, and mineral release into upper layers of the soil has been called the Goldschmidt or Vernadskii principle (see Malyuga, 1964, for full discussion) and it is the reason why humus layers of the soil are often useful in biogeochemical prospecting (see Section VIIIA). Generally metal ions complexed to organic matter are unavailable to plants, so that the effective metal concentration of the soil is reduced.

Schwickerath (1931) produced evidence that colonization of the mine habitat by some species appeared to depend on the prior presence of others: with increasing colonization new species were distinctly associated with the *Festuca ovina* sward. Similarly Schubert (1954b) showed that colonization by *Silene vulgaris* var. *humilis* was concentrated on or near the *Minuartia verna* swards. It cannot be deduced whether such associations are due to a genuine alleviation of available metal level, or whether both species colonize patches of slightly lower toxicity or patches which are ecologically less adverse in some other way.

Schwickerath (1931), Schubert (1954b) and Ernst (1965b) recognize areas of differing colonization and assume that they represent stages of a succession. But the data can equally well be interpreted as a static situation in which a varied set of environments, and not successional processes, are the cause of the different communities. If a truly successional situation is assumed it is still not clear whether it is *autogenic*, in that the plants themselves are alleviating the toxicity, or *allogenic*, in that soil toxicity is being lowered by weathering. Schwickerath (1931) and Ernst (1965b) both show that the level of zinc is lower in the areas of increasing colonization. This suggests that either there has been allogenic succession, or more likely that the extent of colonization is related to the initial zinc content of the soil.

5. *Other factors*

Many metal-contaminated areas are created by mining activity. Since the history of mining in any given area is often well documented it might seem possible to look at the colonization of mine areas in relation to the age of the mine. The most extensive investigation has been by Schubert (1954b), who looked at copper-mine tips in Germany. He studied four main periods of mining between 1200 and the present day, but no clear-cut succession could be seen from one period to another. Differences between the periods could be explained by the various shapes of the tips determining exposure, aspect, and weathering. In the fourth period, more efficient extraction has lowered the copper content and therefore made comparison even more difficult. Antonovics (1966) briefly investigated the relationship between the age of lead mines in Wales and their colonization by *Agrostis tenuis*, but again no significant relationship was found. The classification of mine tips by age not only disregards other ecological factors, but colonization is probably quite rapid and in the order of twenty to thirty years so that historical data provide little useful evidence. The period involved is nevertheless sufficiently long to have eluded direct observation.

Ernst (1969b) has documented by pollen analysis *Armeria* and *Minuartia* on a copper-contaminated moor in Wales. By broad correlative analysis of the vegetation he considers the species to date back at least to the twelfth century AD, when both were at their peak. They were drastically reduced as a result of removing the copper turf around AD 1800. Such studies show the potential age of some populations. But some must be much older than this, perhaps dating back to the post glacial, since ore bodies could well have been exposed by glaciation and caused areas of contamination from that period onwards. In the Congo (Duvigneaud and Denaeyer-de Smet, 1963), Rhodesia (Wild, 1968, 1970) and Australia (Nicolls *et al.*, 1965) it is clear that ore bodies can have outstanding effects on vegetation without the occurrence of any mining activity (see Section VIIIA).

Another important factor in colonization is the proximity of available species. Antonovics (1966) showed that the density of populations of *Agrostis tenuis* on lead mines was greater when the populations were nearer the edge of the mine, nearer the original seed and pollen source. Although this study was inadequate in that metal concentrations in the soil were not investigated, a similar relationship was found by Schubert (1954b), who showed that *Minuartia verna* was found on tips downwind from an area where it was in great abundance. Although the evidence is again rather circumstantial, it does show the possibility that a source of suitable colonists might determine the composition of mine communities.

6. *Conclusion*

It is clear that critical evidence on the relative importance of heavy metals and other factors on the colonization of mine soils is lacking. Even though many ecologists have attempted to gauge the factors in some detail, many of their methods and assumptions will not hold up to critical examination. The situation is not peculiar to the study of metal-tolerant plants since too often factors affecting plant distribution are obtained on the basis of broad correlation principles. The main fault in the methodology used is that local distribution and habitat preference are used to define the physiological amplitude of the species involved, the successional changes that have occurred, the competitive influence of other species and so on. The arguments are then taken full circle and the factors that have been defined by distribution are then used to explain the distribution. The way out of the vicious circle is not only to define the environmental conditions more and more precisely, but to combine such studies with experimental work both in the field and in the greenhouse.

The most critical factor of all which has been rarely considered by ecologists is metal tolerance itself. Evidence will be presented in Section V that the ability to colonize mine areas requires the ability to possess or evolve tolerance: plant distribution on metal-contaminated areas has an evolutionary component. For example, the classification of plants into metallophytes, pseudometallophytes, etc. (Lambinon and Auquier, 1964) outlined earlier, may reflect not simply an ability to *tolerate* high contamination but also an ability to *evolve* such tolerance.

The crucial ecological problem is therefore whether the presence of species on metal-contaminated areas is always dependent on the evolution of tolerance or whether in some cases there are species which possess metal tolerance throughout their range; and *per contra* whether the absence of species from metal-contaminated areas is due to their inability to evolve tolerance. The evidence on this will be discussed in Section VB.

The problems of adaptation to this distinctive environment serve to highlight problems of methodology. An enlightened examination of the vegetation on metal-contaminated areas could be very instructive.

III. Geographical Investigations

Metal-contaminated areas, either naturally occurring or man-made, are very widely distributed. They also, as has been previously shown, carry a very distinct type of vegetation which often contains species largely or wholly restricted to such areas. The geographical distribution of these species is therefore highly disjunct, and this has attracted the attention of several workers.

Schultz (1912) argued that since both *Viola lutea* and *Minuartia verna*, two plants commonly found on contaminated soil, have their main distribution in the alpine regions of Europe (and *Minuartia* also in the Arctic regions and Asiatic mountains) and show elsewhere in Europe a highly disjunct distribution (being largely limited to metal-contaminated areas), they must have been widely distributed throughout Europe at one time when conditions were favourable, in the fourth glacial period. Since then, their distribution has been limited to mine tips where they do not suffer from the effects of competition. This finds support from the evidence of post-glacial distribution for other species with disjunct distributions (Pigott and Walters, 1954; Godwin, 1956). He also considered some other heavy metal plants; he suggested *Thlaspi alpestre* and *Arabis halleri* were also of a formerly wide distribution, but there was no clear pattern for *Armeria halleri* and *Silene vulgaris.*

Similar arguments to those developed by Schultz have been used in connection with a study of *Festuca* spp. growing on contaminated areas in and around Belgium (Auquier, 1964). These fescues are considered to belong to several taxa. None of these taxa is completely restricted to contaminated soils, their relatives being often found in upland areas. However, in Belgium itself these plants are strictly confined to calamine (zinc-contaminated) soils. Auquier infers from this that these taxa represent remains of plants which once had a wide distribution during the arctic-alpine climate of the last ice age. A similar interpretation has been arrived at by Heimans (1960) studying the distribution of the zinc violet, *Viola calaminaria,* but he gives it only his tentative support.

Stebbins (1942), in discussing the interpretation of disjunct distributions, distinguished two types. The "paleo-endemic", which is a species of formerly wide distribution that is now confined to particular areas; and "neo-endemic", which is a species that has originated recently in given areas in response often to peculiar environmental conditions.

The suggestions of Schultz, Auquier, and Heimans are therefore that the species found on metal-contaminated soils are usually palaeoendemics. But the situation is difficult to interpret: a simple palaeoendemic interpretation of the geographical distribution of all the species may often be inaccurate.

Firstly, many of the so-called mine taxa are not distinct and there is no evidence that they are reproductively isolated from their relatives. These taxa could be the products of parallel evolution on different mines from relatives in the neighbouring pastures. This process, which has been shown in grasses (see Section VE), can occur rapidly; and the evolution of tolerance can occur in parallel with evolution of morphological differences. The mine taxa in this case would be neo-endemics. The interpretations of Auquier (1964) are probably erroneous since his taxa of *Festuca* are not highly distinct and could easily have arisen by recent evolution.

Secondly, contaminated soils are commonly associated with man's activities, and the widespread disjunct distribution of many mine plants may simply indicate the efficiency of dispersal by human agencies. This phenomenon could be called "transport endemism". Many of the mines are in mountainous regions and dispersal of alpine members such as *Viola lutea, Minuartia verna* and *Thlaspi alpestre* from upland to lowland mining areas, could give the erroneous impression that these upland plants on lowland areas are remnants from a formerly colder climate.

Thirdly, many metal mines are recent and in no way correlated with natural metal outcrops. Moreover, where outcrops occurred they have almost invariably been destroyed by mining activities. Arguments about plant distribution on contaminated areas which are the result of mining may not involve factors present more than a few hundred years ago.

All these arguments favour strongly the suggestion that most mine species are neo-endemics. Evidence that this has been the case comes from the work of Duvigneaud and Denaeyer-de Smet (1963) on the flora of mainly natural copper outcrops in Katanga. With regard to this flora they suggest that the majority of the cuprophyte species are close relatives of non-cuprophyte species which have an extended distribution and which are all found in the Zambesian region. In other respects the cuprophyte flora consists mainly of sub-species or vicarious varieties with the status of ecotypes and probably a large number of other ecotypes not morphologically differentiated. Its character is that of a young neo-endemic flora having its origin in the same area in which it is found. Some taxa have the character of relics of regional movements corresponding to the displacements which have affected the flora of the Zambesian region.

In this area, therefore, neo-endemism appears to be the rule. In those cases where a neo-endemic hypothesis cannot be applied because the possible source of the neo-endemic is far away, transport by human agency may well be an explanation.

The last and very crucial point is that even if in some instances the evidence does favour the palaeoendemic hypothesis, no plant is known that has an inherent or constitutional resistance to heavy metals throughout its distribution (see Section VB). In other words the plants on mines cannot be regarded as simple "left-overs" from a widespread distribution. The problem of the evolution of tolerance is still there.

The interpretation of the disjunct distribution of metal-tolerant plants therefore poses general questions on the nature of endemism, questions which by no means have yet been fully answered.

IV. Taxonomic Investigations

The difference in morphology of plants growing on metal-contaminated areas from plants in normal areas was noted by Baumann (1885) and Jensch (1894). Numerous workers have since then remarked on this feature of tolerant populations (see Section VF). It is therefore not surprising that taxonomists have attempted to classify these plants as separate taxa. Floras and the literature are strewn with subspecific and varietal epithets referring to mine taxa (e.g. Ernst, 1965a).

The literature on the taxonomy of plants growing on metal-contaminated soils is by far the hardest to review and it is almost impossible to come to any conclusions regarding the validity of the systems proposed for groups that have been studied. The reason for this is partly that the material concerned is in itself difficult to include in any classification system, but mostly because the taxonomy of plants is often as cursory as it is unreliable.

Firstly, there have been very few attempts to grow mine- and non-mine plants under uniform conditions to see if the differences between them (and between different tolerant populations) are inherited: conditions in metal-contaminated areas are so extreme that gross, environmentally induced effects are to be expected and examination of field material may be completely misleading taxonomically. It is interesting in this connection that Malyuga (1964) assumes changes in flower form (*Papaver* spp., *Pulsatilla patens*), flower colour (*Papaver* spp.), and changes in morphology over areas of enhanced toxicity to be the sole result of environmental effects, although there seems no good reason for this conclusion: they may well be genetic.

Secondly, there have only been a few attempts to cross mine and non-mine plants to try and assess the distinctiveness of the "taxa" (see Section VI).

Thirdly, very few taxonomists have considered the possibility that morphological differences (even if inherited) may be the result of direct adaptation to mine habitats and of quite recent origin.

The crux of the matter is that mine habitats are ecologically very distinct and often clearly demarcated from normal habitats. This makes the taxonomic recognition of forms very easy even if they are not very different from other taxa in terms of taxonomic and evolutionary distance. This is particularly so of characters such as dwarfness and hairiness. These forms are therefore given high taxonomic or even specific rank when often they are probably no more than ecotypes.

The methods of experimental taxonomy must be used if any reliable picture is to emerge. So far, not even the most elementary aspects of this approach have been attempted. A few examples will serve to illustrate the state of research in this field.

Auquier (1964) investigated the genus *Festuca*, growing on metal-contaminated soils in and around Belgium and classified the genus into two main species, one of which he divided further into subspecies, varieties and forms. This work can however be criticized from several standpoints. Firstly, material used in this study was collected directly from the field: no attempt was made to grow it under standard conditions. The taxa include the fine delimitations of variety, sub-variety and form; these could easily be environmental modifications. Secondly, even if the differences between the taxa are inherited they could easily have been achieved by parallel evolution on different mines since the characters that were measured (length of floral parts, anatomy of tillers, hairiness, vein number, leaf diameter) could easily have been affected by selection for local adaptation.

The complexities of the situation in the European genera *Silene*, *Minuartia*, *Armeria*, and *Viola* have been reviewed by Schubert (1954a) and Heimans (1960, 1966), but their views on classification of some of the genera do not escape from the controversies and problems of synonyms. Schubert quotes seven names for *Minuartia verna*; these are either synonyms or refer most probably to different ecotypes. Heimans, in his detailed review on the zinc violet (*Viola calaminaria*), quotes eight names of different ranks which are probably synonyms. With regard to the genus *Armeria*, Heimans states "Some taxonomists refer the European representatives of the genus *Armeria* to ten different species, whereas other taxonomists recognize but two". Lefebvre (1967) shows clearly some of the reasons for this confusion: populations of *Armeria* from zinc soils in Belgium are intermediate in form between *A. alpina* and *A. maritima*; the ranges of variation of the populations overlap with both these species, and populations on neighbouring contaminated sites can be very different. Nor do the populations have all the characters of another taxon with species status, *A. elongata* in which they had previously been included.

Duvigneaud (1958, 1959), Duvigneaud and Timperman (1959), and Duvigneaud and Denaeyer-de Smet (1960, 1963) recognize many new species, and new subspecies on metalliferous soils in Katanga. Wild (1968) reports several apparently distinct subspecific taxa although none at the specific level in the Rhodesia copper flora, and recently (1970, pers. comm.) has shown that one taxon previously considered to be *Dicoma macrocephala* is a new and distinct species. Again, it is difficult to assess the validity of many of the categories since the observations are made on material taken directly from the field. However, they are often well aware of the arbitrary nature of their taxonomic units.

There have been a few cytological studies of mine plants, and two examples quoted by Heimans (1960) are of interest. Firstly, his own work on the zinc violet showed that the chromosome number was $2n = 52$ (even though he and other workers had previously recorded it as

$2n = 48$) which differed from that of its supposedly nearest relative, *Viola lutea*. Heimans therefore, and on other evidence, concluded greater affinity between *Viola calaminaria* and *Viola alpestris* ($2n = 36$). He also mentions the work of Duyvendak (no reference or actual chromosome numbers given) which showed that *Festuca ovina* in the zinc area differs from all other strains found in the Netherlands, but the chromosome number agrees with those of strains found in the mountainous regions outside the Netherlands. However, Wilkins (1960) could not relate the ploidy level of *Festuca ovina* to the metal tolerance or to the relative occurrence of the two ploidy types on contaminated and uncontaminated soils. No differences in chromosome number of tolerant and non-tolerant *Agrostis tenuis* were found by Jowett (1959a).

This work on chromosome numbers is interesting in that it suggests that perhaps some of the mine races have been formed by a process similar to catastrophic selection (involving drastic chromosomal adjustments of a specially adapted type, leading to reproductive isolation) described by Lewis (1962). Further studies of chromosome numbers of tolerant and non-tolerant plants could prove rewarding from both the taxonomic and evolutionary viewpoints.

The taxonomy of plants on contaminated areas needs a thorough experimental review. Speculation on the basis of present literature is probably not rewarding unless done in conjunction with experimental work. As a field of research it could again have great relevance to taxonomic methodology, especially with regard to endemic and specialized forms.

V. Evolution of Metal Tolerance

In the absence of any prior knowledge the existence of plants on metal-contaminated soils would immediately raise the question of whether these plants belonged to species that were for some reason inherently tolerant to metals, or whether they were races that had evolved a special tolerance not possessed by the remainder of the species.

As will be shown there is much literature demonstrating the existence of such races and this has not only exposed a new facet that must be taken into account when explaining plant distribution, but also has produced the best available example of natural selection in plants. Heavy metal tolerance in plants is probably an example of more powerful evolution in action than industrial melanism in moths.

A. Evidence for Races Tolerant to Metals

This section considers the evidence that the ability to colonize contaminated areas depends on the ability to evolve races tolerant to heavy metals.

1. Naturally occurring races on contaminated soils

The first comparative study of mine and non-mine populations is that of Prat (1934). Seed of *Melandrium silvestre* from a copper mine tip grew far better in soils with the higher concentrations of added copper carbonate than seed of plants from uncontaminated soil; at the highest concentrations the plants from uncontaminated areas died in the seedling stage. Prat considered the increased resistance of plants growing on copper mines to be the result of natural selection.

This investigation was the only one of its kind until the 1950's, when studies of mine and non-mine populations were resumed independently in Great Britain and Germany.

In Great Britain, Bradshaw (1952) reported populations of *Agrostis tenuis* tolerant to lead and zinc soils; plants from pasture soils did not survive on mine soil. Following this, Wilkins (1957, 1960) developed a rooting technique which showed *Festuca ovina* to have races tolerant to lead. The technique involved comparison of root growth in solutions with and without added metal salts, and this technique or its modifications (Gregory and Bradshaw, 1965; Holmes, 1965; Jowett, 1958; McNeilly, 1965; Gadgil, 1969) has been used to establish and quantify the tolerance of races of a wide range of species found on metal mine soils by a wide variety of authors (Table V). Work in Germany has also produced an impressive list of plants showing tolerant races. Schwanitz and Hahn (1954a, b) recording survival in water culture with different amounts of zinc showed several species to be tolerant if taken from zinc-contaminated soil, but non-tolerant if taken from normal soil. Repp (1963) and Gries (1966) measuring the cellular resistance to metal ions, by finding the concentration that produced death of epidermal cells, established tolerant races in yet other species. The results of work on races tolerant to various metals is summarized in Table V. Baumeister (1954), Baumeister and Burghardt (1956) and Wachsmann (1961) recording rate of photosynthesis and general growth, and Bröker (1963) recording dry weight in water culture, have all confirmed that *Silene inflata* forms races tolerant to zinc and to copper. Url (1956) showed that species of mosses from copper-contaminated regions had a far higher resistance to copper than species from normal soils. This work was not, however, extended to the intraspecific level.

From these results it is clear that the existence of metal-tolerant races is a common phenomenon. Many of the species can develop resistance to more than one metal, either in separate races or simultaneously in the same race if the soil is contaminated with two metals (Jowett, 1958; Gregory and Bradshaw, 1965). Generally speaking, the tolerance is metal specific: tolerance to one metal does not automatically confer

TABLE V

Species showing races tolerant to heavy metals

Species	Metal	Author
Melandrium silvestre	Cu	Prat, 1934
Taraxacum officinale	Cu	Repp, 1963
Tussilago farfara	Cu	Repp, 1963
Mimulus guttatus	Cu	Allen and Sheppard, 1971
Agrostis tenuis	Cu, Pb, Ni, Zn, Cu+Ni, Pb+Cu, Zn+Pb	Gregory and Bradshaw, 1965
	Cu	McNeilly, 1968; McNeilly and Bradshaw, 1968
	Pb	Jowett, 1958, 1964; Jain and Bradshaw, 1966
	Pb, Zn	Bradshaw, 1952
Rumex acetosa	Cu	Coackley and Dawson, 1966
	Zn	Spilling and Thomas, 1964; Schwanitz and Hahn, 1954b
Festuca ovina	Pb	Wilkins, 1957, 1960
	Zn	Gregory and Bradshaw, 1965
Agrostis stolonifera	Pb	Jowett, 1958
	Zn	Gregory and Bradshaw, 1965; Archer, 1964
Agrostis canina	Pb	Craig, 1970
	Zn	Gregory and Bradshaw, 1965
Viola lutea	Zn	Schwanitz and Hahn, 1954a
Alsine (= *Minuartia*) *verna*	Zn	Schwanitz and Hahn, 1954a; Humphreys and Farnworth, 1964
Silene vulgaris (=*inflata*=*cucubalus*)	Zn	Schwanitz and Hahn, 1954a; Baumeister, 1954; Baumeister and Burghardt, 1956; Bröker, 1963; Gries, 1966; Ernst, 1968d
	Cu, Zn	Wachsmann, 1961
Plantago lanceolata	Zn	Schwanitz and Hahn, 1954b; Williams and Morgan, 1964
Linum catharticum	Zn	Schwanitz and Hahn, 1954b
Campanula rotundifolia	Zn	Schwanitz and Hahn, 1954b
Agrostis tenuis×*stolonifera*	Zn	Gregory and Bradshaw, 1965; Archer, 1964
Festuca rubra	Zn	Gregory and Bradshaw, 1965;
Holcus lanatus	Zn	Jenkins and Winfield, 1964; Antonovics, 1966
Anthoxanthum odoratum	Zn	Gregory and Bradshaw, 1965; Putwain, 1963; Antonovics, 1966
Thlaspi alpestre	Zn	Ernst, 1968d
Armeria maritima	Zn	Lefebvre, 1968

tolerance to another (see Section VIIc). Moreover, the level of tolerance developed is related to the amount of metal in the soil. Thus Wilkins (1960) demonstrated a correlation between the degree of lead tolerance of *Festuca ovina* and amount of extractable lead in the soil. Similar relationships between degree of tolerance and amount of metal in the soils have been demonstrated in *Agrostis tenuis* on lead mines by Jowett (1964), on copper mines by Gregory and Bradshaw (1965) and McNeilly and Bradshaw (1968), and on zinc mines by Gregory and Bradshaw (1965) and Turner (1967, 1969).

2. Metal tolerance to sprays and toxicants

In higher plants, there is no evidence of the evolution of tolerance to fungicide and pesticide residues, almost certainly because metallic compounds have not been employed directly on plants in toxic quantities. However, in certain habitats toxic levels of metal are being approached: roadsides are frequently contaminated with lead, and the soil of fruit orchards and vineyards contain large amounts of lead, copper and mercury (see Section I for references). Metal tolerance may well be evolving in such situations.

In lower organisms there is some evidence that races tolerant to metal toxicants have evolved. Toxic paints involving compounds of copper and mercury are used as antifouling compounds on ships. In those cases where algal growth has occurred due to the presence of *Ectocarpus* spp., the populations have been shown to possess tolerance to copper many times greater than in normal populations (Russell and Morris, 1970). A similar tolerance to mercury has been demonstrated in *Enteromorpha* spp. (Hodgson, 1969).

In fungi, there is surprisingly little evidence of tolerant races. Ashida (1965), reviewing the adaptation of fungi to metal-containing toxicants, mentions only two instances, both cases of adaptation to Bordeaux mixtures. Taylor (1953) reports that spores of *Physalophora obtusa* from orchards sprayed with Bordeaux mixture are more resistant to copper than spores of the same species from unsprayed orchards. Horsfall (1956) suggests that the reduction in effectiveness of Bordeaux mixture in controlling potato blight might be due to the evolution of resistance to it by *Phytophthora infestans*.

More recently there have been several reports of resistance of *Pyrenophora avenae* to mercury-containing fungicides (Noble *et al.*, 1966; Malone, 1968; Old, 1968; Sheridan *et al.*, 1968; Bainbridge, 1969; Greenaway and Cowan, 1970). There is considerable evidence that such resistance is spreading (Malone, 1968; Old, 1968). Soil fungi of the genera *Penicillium* and *Aspergillus* may also develop resistance to mercury fungicide residues in the soil (Spanis *et al.*, 1962). Young (1961) found copper-tolerant races of several brown rot fungi regularly associated with copper fungicide failures on fence posts. This was confirmed by Da Costa and Kerruish (1964) who showed that most brown rot fungi were highly tolerant to copper sulphate and chromate. The most tolerant species, *Poria villantii*, grew on agar containing 8% copper sulphate.

There is therefore evidence that the evolution of fungi resistant to metal containing toxicants is possible, but such that examples are relatively rare. This may in part be because the resistant strains are less able to compete with virulent strains in the absence of the fungicide (e.g. poor vitality of copper-resistant *Phytophthora infestans* (D'Yakov, 1963a, b), or simply that the evolution of tolerance to metal toxicants is not a very common phenomenon. The former explanation seems more likely since there is considerable evidence for tolerance to metals in laboratory strains (see Section VA3). However, there is clearly insufficient evidence to come to any firm conclusions.

Bacteria resistant to metals have been isolated from various sources. Chatterjee *et al.* (1968) record mercury-resistant strains of *Staphylococcus pyogenes* and note that their occurrence is more common in hospitals than in the general community. Mercury resistance was common in strains responsible for hospital cross-infections and frequently also had resistance to several antibiotics.

A similar association between mercury resistance and drug resistance (to tetracyclines) was found in *Staphylococcus aureus* (Williams, 1967). Novick and Roth (1968) show that separate factors giving up to a hundredfold resistance to As, Pb, Cd, Hg, Bi, Sb and Zn are associated on a non-nuclear plasmid carrying a separate factor for *Penicillin* resistance.

Mercury-resistant strains of bacteria have been isolated from soils contaminated with mercury-containing fungicides by Tonomura, Nakagami, Futai, and Maeda (1968) (*Pseudomonas* sp.), and by Spanis, Munnecke and Solberg (1962) (*Bacillus* spp).

3. *Metal tolerance in laboratory strains*

Under laboratory conditions the evolution of metal-tolerant races appears to take place fairly frequently and apparently more readily than in nature. Much of the work on laboratory strains of filamentous fungi and yeasts has been reviewed by Ashida (1965) and this section will deal with other groups and subsequent papers on fungi. Both Ashida (1965) and Turner (1969) contain tables showing the tolerances recorded for fungi and a few other micro-organisms.

Relatively few studies have attempted to select for resistance to metal ions in bacteria, although this should be possible since such races often occur naturally (see Sections IIA4, VA2, and VIIIB).

Several recent papers (or papers not mentioned by Ashida, 1965) have been concerned with selection for metal tolerance in fungi. D'Yakov (1963a, b) showed slow adaptation to copper in *Phytophthora infestans* by successive transfer between tuber discs sprayed with copper arsenate solution. The resulting strains were (after seventeen transfers) three times more resistant than the controls, but the spores germinated poorly on transfer to normal medium and the tolerance was lost after two such transfers.

Antoine (1965) paralleled the earlier studies of Ashida and co-workers, and showed the existence of copper-tolerant races of yeast.

The work on natural selection for metal tolerance on mine tips has recently been paralleled by attempts to select artificially for tolerance to toxic soils. Seed of *Agrostis tenuis* from normal populations can be screened to produce in one generation genotypes tolerant to copper and to zinc, whereas other species of grass not normally found on contaminated soils do not produce tolerant races when screened in the same way (see Section VE for full references). However, varietal differences in the tolerance of soybeans to excess quantities of zinc (Earley, 1943) suggest that gradual selection for tolerance may be successful in species not normally colonizing mine soils.

B. THE OCCURRENCE OF TOLERANCE IN DIFFERENT FAMILIES AND SPECIES

The evolution of tolerant races appears to be such a common phenomenon that it is pertinent first of all to ask if there is any evidence that some species are inherently tolerant to metals (even when not growing on metal-contaminated soil) and are thereby able to colonize contaminated areas. The only work which suggests that inherent (or constitutional) tolerance may be important in colonizing metal mines is that of Repp (1963). She showed that the cellular resistance of *Silene vulgaris* (= *cucubalus*, *inflata*) from normal soil was just as high as the resistance of *S. vulgaris* from mine soils. This finding is however in contrast to the whole plant investigations of Bröker (1963) and Schwanitz and Hahn (1954a), and was quite clearly contradicted by the work of Gries (1966) where normal *Silene cucubalus* was shown to be non-resistant by a similar technique. So far, therefore, there is no evidence that a species has constitutional tolerance to heavy metals: evolution has always occurred when mine habitats are colonized.

Because of the work of particular people, it is easy to get the impression that tolerance is restricted to certain families and genera, such as the *Caryophyllaceae*, or the genera *Agrostis* and *Festuca*. But Table V shows that the more work that is done the less this seems to hold.

Good examples of tolerance newly found in separate families are *Mimulus guttatus*, *Thlaspi alpestre* and *Armeria maritima*.

If we assume, as seems reasonable, that all species occurring on high concentrations of metals have evolved tolerant populations, then there is much evidence, particularly that of Wild (1968, 1970), which shows that tolerance is very widely distributed indeed, and is to be found in families as widely separated as Cyperaceae, Compositae, Orchidaceae and Ranunculaceae. The greatest number of tolerant species is found in the Leguminosae and the Gramineae. But these of course happen also to be among the largest families in the world's angiosperm flora with correspondingly greater possibilities of evolution of tolerant species.

The special characteristics of species that are able to evolve tolerance to more than a single metal appears from the restricted evidence in Table V to be equally widely distributed, occurring in species as unrelated as *Silene vulgaris* and *Agrostis tenuis*. The evidence of species occurrence on toxic soil supports this. Many of the species in Rhodesia that occur on copper soils occur also on nickel (Wild, 1970).

However, it is interesting that while some species are able to evolve tolerance, others seem unable to do so even though they are abundant in the vicinity of metal-contaminated soils. Why this should be so is not clear. Species (not normally associated with metal-contaminated soils) have been shown to differ in their tolerance to metals (e.g. Earley, 1943; Biebl, 1947a, b, 1950) or metal uptake (e.g. Gorsline *et al.*, 1964), but no one has critically compared these aspects of plants that can and those that cannot evolve tolerance. However, Khan (1969) has preliminary evidence from a comparison of five species which are known to evolve tolerance and five species which seem not to, that the former possess genes for tolerance in normal populations while the latter do not. Another possibility is that a certain characteristic may pre-adapt certain species to colonize toxic areas: mine habitats require adaptations other than metal tolerance (see Section IIc2) and the possession of these characters (e.g. tolerance to low nutrients) would assist the development of tolerant races. Otherwise, the requirement of multiple adaptation to new features of the environment is considerably more difficult. Simultaneous selection for tolerance to two metals is considerably less successful than selection for one (Khan, 1969; see also Section VE). A paradigm for the idea that certain characters may pre-adapt populations to metals comes from the work of Horn and Wilkie (1966) who show that a respiratory mutant of yeast is advantageous to a cobalt medium.

C. THE "NEED" FOR METALS BY TOLERANT PLANTS

Several workers (e.g. Schatz, 1955) have suggested that tolerant plants may have a positive need for metals and for this reason are restricted to contaminated areas. Work on tolerant races shows that this is not the case: tolerant plants grow well in normal soil. However, there is some evidence that tolerant plants are stimulated in their growth by levels of metal considerably above the normal micro-nutrient levels, and more so than non-tolerant plants.

Baumeister (1954) and Baumeister and Burghardt (1956) grew *Silene vulgaris* at different levels of zinc in soil, sand-culture, and water-culture. At intermediate levels of zinc the tolerant type had a greater fresh weight, rate of photosynthesis, and chlorophyll content than at low levels of zinc. The effect was not so marked in the non-tolerant type.

Lefebvre (1969) showed that plants of *Armeria* from zinc-contaminated soil were much stimulated in their growth on normal soil by the addition of zinc salts.

Further evidence comes from authors who have found indices of tolerance greater than 100% (more root growth in metal than without metal). These are listed in Table VI.

TABLE VI

Species showing an index of tolerance indicative of root growth being stimulated by presence of metal ions in the testing solution

Species	Metal	Author
Agrostis tenuis	Cu	McNeilly, 1968
	Pb	Jowett, 1964
	Pb, Cu, Zn	Barker, 1967
Mimulus guttatus	Cu	Allen and Sheppard, 1971
Anthoxanthum odoratum	Zn	Putwain, 1963; Antonovics, 1966; Gadgil, 1969
Holcus lanatus	Zn	Jenkins and Winfield, 1964
Armeria maritima	Zn	Lefebvre, 1969

Although Ernst (1965b) did not use control non-tolerant populations, five out of six species from zinc soils produced a greater percentage germination and rate of germination in 50 ppm zinc than in 1 ppm. *Epilobium angustifolium*, a species from a normal area, did not show this effect.

There is therefore considerable evidence that tolerant plants are stimulated in their growth by small amounts of metal. This could be interpreted as a definite need for metal in these plants, but a fairer interpretation is probably that, because of the efficiency of the tolerance mechanism in inactivating the metals, the external trace element requirement is rather higher than the norm.

D. THE COMPETITIVE PERFORMANCE OF TOLERANT PLANTS

The apparent restriction of metal-tolerant genotypes on contaminated areas, despite their normal growth on non-contaminated soil, has led to the speculation that these plants are largely restricted to mine soils because they are competitively inferior to normal plants. This idea is supported by Kruckeberg (1954), who showed that serpentine species did not survive when grown in competition with their non-serpentine counterparts.

Putwain (1963) and Antonovics (1966) have shown that zinc-tolerant *Anthoxanthum odoratum* is competitively inferior to non-tolerant when grown on non-contaminated soil. Putwain (see Jain and Bradshaw, 1966) estimates that the selection pressure against the tolerant type on normal soils is 0·3. Antonovics (1966) showed that yield of tolerant type was progressively reduced in competition, whereas that of non-tolerant increased: after nine months the selection pressure was at 0·35 against the tolerant type.

The situation in *Agrostis tenuis* is rather different. McNeilly (1965) not only failed to find any differences in fitness between copper-tolerant and ordinary when grown as spaced plants, but also found only just significant differences when these were grown in competition. From the latter experiment the selection pressure against the tolerant type is only 0·05. Earlier work of Bradshaw (see Jain and Bradshaw, 1966) suggested that selection against lead tolerance could be as high as 0·4, but it is possible that other characters besides metal tolerance were involved.

Further competition experiments between plants in sand culture (McNeilly, 1970) have shown that the outcome of competition between tolerant and non-tolerant plants depends on phosphate level: at high phosphate levels the non-tolerant plants are competitively superior, whereas the converse is true at low phosphate levels.

Lefebvre (1970a) studying the competition of tolerant and non-tolerant ecotypes of *Agrostis tenuis*, *Anthoxanthum odoratum* and *Plantago lanceolata*, each grown with a normal herbage grass, *Lolium perenne*, showed that the tolerant ecotypes were competitively inferior to the non-tolerant.

E. THE PROCESS OF EVOLUTION

The first discovery by Prat (1934) of plant populations on metal mines tolerant to heavy metals in the soil led to his postulating that metal tolerance had arisen by natural selection.

The existence of a character that was clearly adaptive to the environment in which it was found, and the discovery by Wilkins (1957) of a rooting technique of measuring metal tolerance, has been exploited by various workers to provide one of the best documented cases of evolution in action. The account which follows is a brief summary of this work mainly on *Agrostis tenuis*.

As has already been mentioned, waste-heaps from mining activities carry tolerant populations of *Agrostis tenuis*, with specific tolerances corresponding to the type of metal in the soil (Gregory and Bradshaw, 1965). When these populations are sampled as seed it is found that they normally have a lower mean tolerance and greater variance than populations sampled as adults (Antonovics, 1966; McNeilly and Bradshaw, 1968; McNeilly, 1968). This is the most direct evidence that in natural conditions there is selection acting to preserve the high tolerance of the adult populations. Selection pressures acting to maintain the populations have been calculated (Antonovics, 1966) and are in the order of 0·3–0·7. Experimental studies have shown that most of the selection occurs at the seedling stage: if non-tolerant seed is sown on contaminated soil, most of the seed germinates (there is a slight inhibition) but the seedlings fail to root, and die before they can produce many leaves (McNeilly, 1968): this suggests that selection pressures can be up to 1·0.

Tolerant populations are normally surrounded by normal non-tolerant populations on ordinary pasture. Gene flow between the two populations is possible since the areas of contamination are usually small or there is a sharp boundary between the two populations at the edge of the mine (McNeilly, 1968). Even on large diffuse mines the background rain of pollen can reduce the level of tolerance of the seed on the mine (McNeilly and Bradshaw, 1968). The existence of gene exchange has been demonstrated by a comparison of seed populations collected in the wild and seed populations produced by adults brought back to the experimental greenhouse and grown away from non-tolerant material. The seed from natural populations has normally a lower tolerance than seed grown in isolation, showing that there is dilution of tolerance by the pollen from the surrounding non-tolerant population (McNeilly, 1968).

However, investigations of tolerance across mine boundaries show that tolerance changes abruptly at the boundary over a distance of a few feet (Jain and Bradshaw, 1966; McNeilly, 1968; Antonovics and Bradshaw, 1970). It seems puzzling that firstly, population differentiation could occur over such short distances, and secondly that there are very few tolerant individuals in adjacent pastures. The former is explicable on the basis that the selection pressures involved are sufficiently strong to maintain the populations in the face of gene flow. This process was examined theoretically using a computer simulation by Jain and Bradshaw (1966), who showed such differentiation to be not only possible but very probable. The absence of tolerant individuals in pasture populations could be explained on the basis of their poorer competitive ability (see Section VD). But as has been shown, the fitness of mine plants on ordinary soil, while less than that of pasture plants, is still relatively high; in which case the selection pressures (even if they were cumulative in perennial grasses) against tolerance in pasture would be much smaller than selection for tolerance on the mine. From this it had to be argued that to fit the findings of the computer model, gene flow on the transects that had been examined must be quite low, or selection high.

The situation was illuminated considerably by the discovery of a copper mine in a U-shaped valley with the prevailing winds in one direction, where a study was made of transects across the mine boundaries in a cross-wind and down-wind direction (McNeilly, 1968). The cross-wind transect showed a sharp cline at the mine boundary whereas in the down-wind transect the tolerance decreased gradually over a long distance (100 m) into the pasture. Moreover, whereas on the mine the seed had a lower tolerance than the parents, in the pasture down-wind from the mine the seed had a greater tolerance than the parents: clearly there is selection against tolerance in the pasture. This is excellent evidence for the occurrence of natural selection and shows clearly the interaction between selection and different amounts of gene flow in determining the pattern of differentiation.

Antonovics and Bradshaw (1970) studied the way in which several characters in addition to tolerance changed across an abrupt zinc/lead mine and pasture boundary. Different clinal patterns were observed for the various characters even though the characters were all measured on the same set of plants. This was interpreted to be a consequence of different selection pressures for the different characters interacting with gene flow which must have been identical for all characters.

In addition to showing differences between the populations in a wide range of characters, this work showed that even within mine and within pasture populations there were differences between the various sites. Mine and pasture habitats were themselves heterogeneous and this led to further differentiation within the populations. There was therefore evidence of disruptive selection maintaining differences within and between the mine and pasture populations.

All this work has shown that there can be differentiation in the face of gene flow, yet that gene flow is not without effect and interacts with selection to produce various clinal patterns. It is also clear that this gene flow must be having a deleterious effect in that it is hindering adaptation to the two environments diluting tolerance on the mine and leading to the production of ill-adapted tolerant seed in the pasture. The magnitude of such a gene-flow load was examined in a computer model by Antonovics (1968b), who showed that it was greater if the selection pressure and the gene flow were greater. In addition, the gene flow load was greater the lower the frequency of the favoured gene; it also depended on the degree of dominance of the favoured gene, being greater if the favoured gene was recessive.

It might be expected that mine and pasture populations had developed mechanisms to reduce the harmful effects of gene flow and a study was therefore made of possible breeding barriers (Antonovics, 1968a; McNeilly and Antonovics, 1968). There was no clear-cut incompatibility between tolerant and non-tolerant populations: crosses between populations were just as successful as crosses within populations. Differences in flowering time were observed between tolerant and non-tolerant populations in the direction of earlier flowering in mine populations (giving approximately 25% isolation) and evidence was presented that sometimes this is an adaptation to gene flow. The self-fertility of mine populations was shown to be considerably greater than the self-fertility of pasture populations. This change in the breeding system could not be clearly attributed to any one factor but a brief survey suggested that this too might be an adaptation to prevent the dilution of tolerance by flow of non-tolerant genes from neighbouring populations. The process of evolution in areas subjected to gene flow was studied by a parallel investigation using computer simulation (Antonovics, 1968b), where it was confirmed that the changes outlined above were quite possible and other changes, such as evolution of dominance in the face of gene flow, were predicted.

The evolution of metal tolerance therefore also sheds light on the mechanism of speciation. Geographical isolation is clearly not a prerequisite for divergence: but lack of isolation can encourage the production of breeding barriers (see Smith, 1966, for theoretical discussion).

Further indirect evidence for changes in characters that may effect changes in the breeding system (or isolation from non-tolerant populations) comes from the work of Malyuga (1964). The reports are that *Papaver macrostomium* has notched petals when growing over zinc-contaminated areas, that *Papaver commutatum* has black bands on the petals when they are growing over copper and molybdenum areas, and that *Pulsatilla patens* lacks petals on nickel silicates in the Kimpersaikii ore region. Malyuga (1950, 1964) suggests that these are direct environmental effects; but the effects are quite specific and unlike the generalized effects of metal toxicity. This suggests that they are, in fact, the result of evolutionary changes of some kind.

The changes in petal shape and markings may be isolating mechanisms (cf. Levin and Kerster, 1967) and the loss of petals suggests a change to inbreeding. It is interesting in this connection that Duvigneaud (1958) reports that *Protea goetzeana* has a creeping sterile form on toxic soils.

Differences between the corolla markings and colours in different *Gladiolus* species of the *Gladiolus robiliartianus* group have been recorded by Duvigneaud and Denaeyer-de Smet (1963). Each *Gladiolus* species appears on a single or group of adjacent copper hills (outcrops), and is characterized by its own corolla type. They differ in other respects such as the position of the inflorescence and leaf length. Such markings may clearly keep the populations isolated from each other and from related non-tolerant types, but their evolutionary origin is difficult to deduce.

A change towards inbreeding has been reported in calamine populations of *Armeria maritima*: they are much more self-fertile than their maritime counterparts (Lefebvre, 1970b). The tolerant *Armeria* populations are not immediately surrounded by non-tolerant forms and therefore self-fertility cannot be an adaptation to reduce the harmful effects of gene flow. It appears that the self-fertility originated and is maintained as a result of the difficulties of establishment of self-sterile *Armeria* on new mines and on the continually changing terrain of calamine areas. This is another example of the far-reaching evolutionary changes in metal-tolerant plants.

The evolution of metal tolerance is of additional interest since it appears to have been extremely rapid. Most mine tips are of relatively recent origin, between 50 and 100 years old (Jones, 1922), yet often carry well-established populations. However, it could be argued that in mining areas there exist long-established pockets of tolerance with these perhaps originally deriving from plants growing in metal outcrops: this is well illustrated by Duvigneaud and Denaeyer-de Smet (1963) and Nicolls *et al.* (1965).

However, tolerant populations exist on contaminated slag heaps which are the result of smelting ore: in the Swansea area of South Wales these are more than fifty miles from the nearest mine (Gregory and Bradshaw, 1965). More convincingly Snaydon (quoted in Bradshaw *et al.*, 1965) showed that plants of *Festuca ovina* and *Agrostis canina* growing under galvanized zinc fences several hundred miles from a mining area were tolerant even though the fences were only erected in 1936.

A new dimension was opened on this problem by the work of Abbot and Misir (quoted in Antonovics, 1966), who showed that it is possible to select for tolerance in *Agrostis tenuis* in one generation. Commercial seed was sown on copper-mine soil alleviated with a small amount of loam and this resulted in the survival of a few plants (1 in 7000).

These few plants, which grew well on the contaminated soil, were planted to normal soil and tested for tolerance. They were found to have a significantly greater tolerance than the original population (Antonovics, 1966). Moreover, the size of the plants on copper soil was significantly related to their tolerance. These studies showed that the evolution of tolerance can take place within one generation by the simple process of screening non-tolerant populations. Walley (1967) confirmed these results; using this technique to pick out tolerant individuals, he showed that seed collected from pasture populations one and a half miles down-wind from a large copper mine produced a significantly greater proportion of tolerant individuals than seed collected from pastures not in the vicinity of the mine. This work illustrated very clearly the long-distance transport of genes by pollen flow.

Khan (1969) not only confirmed these results of Walley but showed that the number of tolerant individuals found in pastures up-wind of the mine was much lower than down-wind. He also made a general study of the problem of evolution of tolerance *de novo* (Bradshaw *et al.*, 1969). He showed that tolerant genotypes occurred in normal populations with a frequency of about 2% for Zn soil and a similar frequency for Cu soil, yet that selection for simultaneous tolerance to two metals was far less successful, suggesting that the occurrence of the tolerances is independent. He was able to demonstrate that the survivors possessed the appropriate tolerances and that the most successful survivors had tolerances approaching that of established tolerant populations. When he selected for zinc tolerance in populations already tolerant to copper, and vice versa, he found the frequency of occurrence was the same as in normal populations lacking tolerance to either metal. Selection for zinc tolerance in a population already tolerant to copper gave survivors that were tolerant to both copper and zinc.

The results on artificial selection for tolerance, and the fact that the species are perennial, might suggest that the process of evolution of tolerance in grasses on metal mines is not a continuing process: tolerant individuals are screened from normal populations in one generation and being perennial persist in subsequent years. However, Antonovics (1966, and subsequent unpublished data) studied by mapping techniques the longevity of *Anthoxanthum odoratum* on a lead/zinc mine, and demonstrated a high rate of population turnover. Only a minute fraction of individuals (3·0%) survived for more than four years. Clearly the evolution of tolerance, and of the other differences between tolerant and pasture populations (see Section VF) is a continuous dynamic process, the characteristics of the mine plants being continually selected for in the face of the diluting effect of gene flow from pasture populations.

The study of the evolution of metal tolerance has added force to the idea that disruptive selection plays an important part in differentiating and possibly eventually completely isolating natural plant populations. Much of the current interest in metal tolerance stems from the light its study can throw on evolution in nature as opposed to the laboratory.

F. EVOLUTION IN ASSOCIATED CHARACTERS

Metal-tolerant plants may differ from normal plants in features other than tolerance. Schwanitz and Hahn (1954a, b) grew tolerant and non-tolerant populations of a range of species under standard conditions and showed that, in general, tolerant plants had smaller flowers, smaller leaves and thinner stems. These differences were reflected in the size of the epidermal cells of the leaves, but using other non-tolerant variants they showed that small cell size did not go hand in hand with tolerance. They concluded that there must be independent selection for morphology as well as tolerance. This conclusion was confirmed in *Silene vulgaris* by Bröker (1963): in F2 progeny of tolerant × non-tolerant crosses, tolerance was not associated with morphology. The first mine population of *Agrostis tenuis* to be examined morphologically (Bradshaw, 1959) showed startling morphological differences from the adjacent pasture population. However, subsequent investigations by Jowett (1964) showed that the mean differences of mine populations from pasture populations was not great.

The data suggest a complex adaptation of each population to local ecological factors independent of tolerance rather than an effect of tolerance on overall size and yield. Only slight differences in morphology between copper-tolerant and non-tolerant plants were detected by McNeilly (1965).

The situation with regard to morphological characters was re-investigated using *Anthoxanthum* by Antonovics and Bradshaw (1970). It was shown that tolerant and non-tolerant individuals from adjacent populations differed in a wide range of characters. Tolerant plants were generally shorter, had a smaller number of flowering tillers, shorter flag leaves, and a lower ratio of flowering/vegetative tillers. The characters, all measured on the same set of plants, formed different clinal patterns across the mine boundary. Gene flow must have been the same for all characters and the results suggested that there are different selection pressures acting on the morphological characters. Within the mine there was no correlation between tolerance and any of these morphological characters, again confirming that selection for them is independent of selection for tolerance.

Observations on other species brought back and grown under uniform conditions (*Rumex acetosa* and *Plantago lanceolata*, unpublished data) confirm that tolerant plants are generally dwarf or prostrate compared with their pasture counterparts.

The differences between tolerant and non-tolerant plants are not confined to morphology. Physiological differences have also been found. Jowett (1959b) and Bradshaw *et al.* (1960) showed that lead-tolerant *Agrostis tenuis* was more tolerant of low calcium and low phosphate, and McNeilly (1965) confirmed the differential phosphate response in copper-tolerant and non-tolerant ecotypes. These differences could be related to the characteristics of the mine soils: they are generally low in phosphate and those studied by Jowett (1959b) were acidic mines deficient in calcium. Barker (1967) showed that *Agrostis tenuis* from a calcareous lead/zinc mine showed good calcium response. Differences between copper-tolerant and non-tolerant plants in response to varying sulphur levels have been shown by Turner (see Antonovics *et al.*, 1967). Non-tolerant plants performed relatively better at low sulphur levels than did copper-tolerant plants. This could again be related to the normally high sulphur content of the mine soil from which the plants were taken.

It is therefore seen that the evolution of tolerance to metals is but one facet to the evolution of plants on abandoned mine workings. There is also natural selection for morphological and physiological characters which adapt the plants to the harsher mine environment. The existence of such changes in associated characters is important in several ways. Firstly, it would be interesting to investigate some of the processes (e.g. gene flow, selection) found in mine populations using these characters as well as tolerance: the course of evolution should be similar. Secondly, morphological differences between tolerant and non-tolerant plants have been the basis of taxonomic distinctions between the two types (Schubert 1954a; Duvigneaud and Denaeyer-de Smet, 1963; Lambinon and Auquier, 1964). It must be noted, however, that in all these studies no attempt was made to grow the plants under standard conditions: the differences recorded could have been purely environmental. Nor was any attempt made to see if there was an absolute correlation between metal tolerance and morphology. Those studies showing differences between plants grown under standard conditions suggest that the taxa that have been recognized by the above workers may either have the status of different ecotypes or imperfectly differentiated local populations.

VI. The Genetics of Heavy Metal Tolerance

There has been very little work done on the genetics of heavy metal tolerance. Wilkins (1960) working on lead tolerance in *Festuca ovina* states that "in spite of the amount of effort devoted to refining the measurement of tolerance, the nature of the genetic mechanism controlling it has not been established with certainty". He nevertheless found that tolerance (whether high or medium range) was dominant and that a major gene with just two alleles was an inadequate model to explain the results. Whether these several alleles were at one locus or at more than one was not established.

Bröker (1963) again found dominant inheritance of zinc tolerance in *Silene vulgaris*. However, the F2 data (from selfing F1's) allowed one to conclude very little since only a few plants per family were tested: it was, therefore, again not decided whether segregation was continuous or discontinuous. These results were confirmed by Gries (1966) using cytoplasmic resistance in the same material as a measure of tolerance.

That many genes are involved in the determination of tolerance is supported by Jowett (1959a), McNeilly (1965), and Antonovics (1966), who showed tolerance to be *not* an all or nothing effect but a continuous variable in natural populations.

Walley (1967) and Khan (1969), discussed in Bradshaw *et al.* (1969), carried out experiments in which they screened normal populations for the occurrence of tolerant individuals. They both found a low frequency of survivors when seed of normal non-tolerant populations was sown on toxic soil (see Section VE). When normal populations were screened for tolerant individuals on soils in which the toxicity was ameliorated to different degrees by the addition of normal soil, the number of survivors varied markedly, increasing with decreased toxicity. At any one level of toxicity survivors and non-survivors could be readily distinguished. This strongly suggests a threshold character determined by several genes.

Antonovics (1966) obtained some evidence that the genetics of copper tolerance in *Agrostis tenuis* might differ on different mines, behaving as a dominant character in crosses from one mine, but as non-dominant in crosses from another mine. However, the number of crosses was too low to allow a firm conclusion on this.

More specific studies by Jowett (1959a) on the genetics of tolerance showed indications that the dominance of lead tolerance in *Agrostis* was variable, but that in general it was partly recessive. Jowett suggested that this could be an artefact of pre-culture conditions which were different in parents and progeny. Nevertheless, his data provide evidence for continuous variation in the character and also considerable segregation, suggesting quite marked heterozygosity of the parents.

Investigations of the genetics of copper tolerance of *Mimulus guttatus* have recently been initiated by Allen and Sheppard (1971). Copper tolerance is readily demonstrable in populations from copper-contaminated situations: populations from normal soils were non-tolerant, except in one case. There was some indication that copper-tolerant populations were also tolerant to zinc, lead and nickel, although these metals were not present in the soils. Tests on F1 progeny of crosses between tolerant and non-tolerant parents showed that there was no maternal effect and that copper tolerance was dominant at low copper concentrations, intermediate at intermediate concentrations, and recessive at high copper concentrations. This can be explained by the assumption of a threshold effect dependent on copper concentration, which would agree with the general conclusions about the expression of tolerance. There was some indication of a copper requirement by tolerant populations at low copper concentration. It was not possible to conclude whether one or several genes were involved.

Similar general evidence that the metal tolerance is inherited comes from the work of Lefebvre (1970a), whose data suggest that the character of zinc tolerance in *Armeria maritima* tends to be dominant.

McNeilly and Bradshaw (1968) compared the copper tolerance of seed and adults of different populations, and found a high correlation ($r = 0{\cdot}983$) between the two. This suggests that the character of tolerance has a high heritability. Further studies of seed obtained from a polycross of tolerant material (McNeilly and Bradshaw, 1968) gave a heritability of 0·7.

Recently Urquhart (1971) has carried out a diallel analysis of lead tolerance in *Festuca ovina*. This has shown that there is additive genetic variation and little maternal effect and that the character must be determined by several genes. Tolerance usually showed complete dominance, but this was not always the case, for in some crosses no dominance was found. This confirms the results of Jowett (1959a), who found that dominance varied between crosses but was inclined at that time to dismiss it as being due to error.

Further evidence of the genetic control of metal tolerance comes from a study of the phenomenon in yeasts. Seno (1962) found that when yeast strains were grown in high copper concentrations they produced two levels of resistance. Both levels were controlled by dominant genes which were very closely linked. Similar results have been obtained by Antoine (1965): here four alleles at one locus were considered to be responsible for copper resistance in yeast. Again the alleles for copper resistance were dominant to the non-resistance allele. It is interesting that evidence was also presented for a remarkable effectiveness of copper ions in mutating the gene for copper resistance to alleles of higher resistance: in this instance mutation appears to be directed. General mutagenic activity of copper ions has been demonstrated by several workers (e.g. Von Rosen, 1964; Bhatia and Narayanan, 1965) and may therefore be important in the evolution of metal tolerance itself. However, there is no evidence for this in higher plants.

So far most of the studies on the genetics of metal tolerance have been cursory and rather inconclusive. Antonovics (1968b) in a computer study of the evolution of metal tolerance on a small mine predicted that the plants would be highly heterozygous and that the evolution of dominance may occur readily. This would explain some of the inconsistencies in the results of Jowett (1959a), Wilkins (1960) and Antonovics (1966), as well as the consistently large amounts of segregation observed in seed from tolerant plants (McNeilly, 1968).

Clearly a full reinvestigation of the genetics of metal tolerance is needed to complement the study of the evolution of tolerance, particularly in grasses on metal mines. This reinvestigation should bear in mind that the level of gene flow into these populations will affect the level of heterozygosity of the plants: it may also promote evolution of dominance (Antonovics, 1968b). Such a study should aim to answer the following questions. What is the dominance of the character? Is it determined by one or many genes? Is it determined differently on different mines? What is the relationship between tolerances to different metals? What is the degree of heterozygosity of the genes for tolerance in the populations?

The genetics of metal tolerance would not only help to complete the evolutionary story, but it could have considerable bearing on the mechanism of tolerance, particularly since such tolerance appears to be highly specific.

VII. Mechanisms of Metal Tolerance

The mechanisms whereby the living organism can grow on apparently contaminated soil may prevent the heavy metals from reaching their sites of toxic action within the plant or they may simply be external factors that prevent the metals from entering the organism. Table VII shows more fully the possible mechanisms whereby heavy metal tolerance may be achieved. These have been adopted and augmented from the proposals of Davis (1958), Ashida (1965) and Ehrlich (1965).

A. External "Tolerance" Mechanisms

The external mechanism of tolerance represents those circumstances which prevent entry of the metal ions. These are not "mechanisms" in the strict sense as they are not under the control of the tolerant organism, but are of considerable ecological importance (see Section IIc).

Firstly, the form of the metal may not be readily soluble in water or if dissolved is rapidly diluted by surrounding water; in both instances

TABLE VII

Possible mechanisms of metal tolerance

A. *External*	B. *Internal*
(i) Form of metal is not directly soluble in water and/or if dissolved then rapidly diluted by surrounding water.	(v) Differential uptake of ions.
(ii) Actual amount of freely diffusable metal ions is small compared to total amount present.	(vi) Removal of metal ions from metabolism by deposition in vacuole.
(iii) Lack of permeability to heavy metals under specific conditions.	(vii) Removal of metal ions from metabolism by pumping from cell.
(iv) Metal ion antagonisms.	(viii) Removal of metal ions from metabolism by rendering into an innocuous form.
	(ix) Excretory mechanisms—removal of "metal storage organ".
	(x) Greater requirement of enzyme systems for metal ions.
	(xi) Alternative metabolic pathway by-passing inhibited site.
	(xii) Increased concentration of metabolite that antagonizes inhibitor.
	(xiii) Increased concentration of enzyme that is inhibited.
	(xiv) Decreased requirement for products of inhibited system.
	(xv) Formation of altered enzyme with decreased affinity for inhibitor or increased relative affinity for substrate compared to the competitive inhibitor.
	(xvi) Decreased permeability of cell or subcellular units to metal ions.
	(xvii) Alteration in protoplasm so that enzymes may function even when toxic metals replace physiological metals.

the effective concentration is low. Ehrlich (1963a), for example, demonstrated that manganese can be released from root nodules by microbial action but that the effective manganese concentration is small due to high water content. It is noticeable that vegetation growth increases considerably in wet places on toxic soils.

The majority of naturally occurring metallic ores are usually found as mixtures of the sulphides which are virtually insoluble: thus the amounts of metal available to plant life are very small.

Temple and Le Roux (1964) showed that colloidal material containing large amounts of metals could be precipitated as the sulphides by sulphate-reducing bacteria. A practical example of this mechanism being used under field conditions has been reported by Lawrence and McCarty (1965). The organisms responsible for digestion of sewage sludge are extremely sensitive to heavy metals (zinc, copper and nickel). Heavy metal toxicity (concentrations may reach 800 ppm) may be prevented by the addition of sulphate ions under anaerobic conditions so that the metals are precipitated as sulphides. Dykeman and De Sousa (1966) showed that in a peat swamp over bedrock material rich in copper, deeply circulating waters dissolve the copper and it percolates up to the surface. The peaty humus at the surface acts as an efficient natural chelating system and toxic quantities of the metal are absorbed and thus immobilized. It is interesting to speculate on the system that originally produced the organic compounds which at present act as chelators: the situation might originally need a tolerant organism to produce organic waste products, thus producing a convenient habitat for succeeding non-tolerant species. Russell (1955) showed that this type of situation can occur in micro-organisms. *Penicillium roqueforti* is apparently tolerant to the phenyl mercuric acetate fungicides used in paper production. *P. roqueforti* absorbs mercury into its mycelium, reducing the amounts of toxic mercury remaining in the medium. This allows growth of subsequent wood-rotting fungi.

A second external mechanism of heavy metal tolerance may be the lack of permeability to heavy metals under specific conditions. Alterations in the environment can produce changes in the permeability properties of the cells or in their uptake characteristics in relation to metals. Hofsten (1962) found *Escherichia coli* to be copper tolerant only under aerobic conditions and Hassall (1962) found *Chlorella vulgaris* tolerant to barium, manganese, nickel, lead and copper under anaerobic conditions. Later work (McBrian and Hassall, 1965) showed that cells under anaerobic conditions absorbed less copper than under aerobic conditions, indicating a difference in uptake potential under varying conditions of aeration. Murayama (1961) found that copper-tolerant yeasts that were respiration deficient, because they lacked part of the normal cytochrome component, were inhibited more strongly by copper under anaerobic conditions. Investigations on a wide range of wood-rotting fungi have demonstrated that copper-tolerant organisms increase their tolerance under conditions of low pH (Young, 1961). A similar situation has been recorded with *Thiobacillus thio-oxidans* and *Azotobacter indicum* (Starkey and Waksman, 1943). Starkey (1964), considering his earlier work, suggested that the ability to develop tolerance under low pH depends on the impermeability of the cells to copper. Lowering the pH to very acid levels would produce, in theory, a net positive charge at the cell surface which would reduce uptake of the metal cations.

Vernon *et al.* (1960), studying a ferrous ion oxidizing bacterium *Ferrobacillus ferro-oxidans*, showed that it had an unusual cytochrome composition that differed from the species that did not oxidize iron. Iron oxidation via the cytochrome oxidase system occurred at the cell surface and the metal entered the cell in large amounts during the microbial breakdown of iron sulphide ores. Ehrlich (1962) suggested that the ability of the same organism to oxidize copper ores was dependent on a similar system. Metal oxidation occurred at the cell surface and the oxidation products were removed by hydrolysis or precipitation. Metal sulphides could thus be regarded simply as external energy sources.

A final external mechanism may be that the effect of toxic ions is reduced by the presence of other ions. Additional ions may reduce the net concentration of toxic ions (i.e. reduce availability simply by competitive effects in solution or by complexing with the toxic ions) or they may interfere in uptake mechanisms by competing for entry sites. Gregory (1965) showed that non-tolerant plants of *Agrostis tenuis* could grow on normally toxic mine soil if a full nutrient culture was provided daily, and toxicity could be markedly reduced if only potassium nitrate was added. Sudzuki-Hills (1963) commented that copper toxicity of mine water was reduced by the presence of other ions. Calcium ameliorates lead toxicity in *Festuca ovina* (Wilkins, 1957) and *Agrostis tenuis* (Jowett, 1964) and also zinc and nickel toxicity in *Agrostis tenuis* (Barker, 1967). Chester (1965) concluded that calcium exerted a protective action against copper poisoning in yeasts. Wilkins (1957) showed the effects to be due to calcium ions independent of the anions supplied. Schmidt *et al.* (1965) found that calcium reduced zinc uptake in barley, and Somers (1963) reported that copper uptake in fungal spores was reduced by calcium. From evidence such as this and related work (e.g. Jefferies *et al.*, 1969) it may be concluded that calcium is important in cell permeability and may exert its influence on metal tolerance in this manner. Calcium may also affect the expression and the type of metal tolerance. Gregory (1965), using calcium nitrate added to the testing solution, found a zinc-tolerant clone of *Agrostis tenuis* to be co-tolerant (defined by Turner and Gregory, 1967) towards nickel and cadmium. Using a modified testing procedure but with no calcium present, Barker (1967) could find no nickel or cadmium co-tolerance in identical plant material.

This action in reducing metal cation uptake is not restricted to calcium, and interference from other metals may occur. For example, copper interferes with zinc uptake in barley (Schmidt *et al.*, 1965). Hoekstra (1964) proposed that certain physical chemical properties that are similar between ions will explain these antagonisms. Vose and Randall (1962), showed that manganese and aluminium-resistant varieties of ryegrass (*Lolium perenne*) are characterized by having a low root cation exchange capacity when compared to susceptible varieties. The authors suggested that a low cation exchange capacity will permit entry of monovalent cations whilst rejecting divalent metal cations; tolerance to heavy metals would therefore be achieved.

B. UPTAKE STUDIES

Physiological investigations on plants growing on metal-contaminated soils have in the past been largely limited to measuring the levels of the various metals in the plants, and in the soil on which they grow. These studies throw some light on the tolerance mechanism since they indicate whether plants are able to tolerate metals by some exclusion mechanism, or whether the mechanism is internal. Almost invariably these studies deal with uptake into the shoot. Uptake by roots has rarely been studied in nature since it is difficult to remove contaminated soil particles from the root system.

1. Zinc

The occurrence of zinc in plants on contaminated soils has been established since the last century (Baumann, 1885; Jensch, 1894). In all these studies, the exact quantities of zinc found are probably not very reliable because of the dangers of contamination from chemicals, glassware and other apparatus. However, all these workers found extremely high values of zinc in plants on zinc soil (and low values on plants on normal soil). Since then the high level of zinc in plants on zinc soils has been consistently, if often briefly, confirmed (e.g. Linstow, 1929; Bertrand and Andreitcheva, 1933; and Dorn, 1937).

There have, however, been more extensive studies which have thrown considerable light on the characteristics of zinc uptake on contaminated soils. The following characteristics have been observed by Robinson *et al.* (1947); Maquinay and Ramaut (1960); Maquinay *et al.* (1961); Lambinon (1964); Lambinon *et al.* (1964); Ernst (1965a); Nicolls *et al.* (1965), and Baumeister (1967).

1. Different species, even though they come from the same contaminated area, differ in the degree to which they take up zinc.

Average values for the more common members of calamine soils are as follows (Ernst, 1965b):

	Zinc content in ppm dry wt	No. of sites investigated
Thlaspi alpestre ssp. *calaminaria*	7757	7
Armeria maritima ssp. *halleri*	3328	5
Minuartia verna ssp. *hercynica*	3007	17
Silene cucubalus (=*vulgaris*) var. *humilis*	1719	27
Armeria maritima ssp. *calaminaria*	1895	2
Viola calaminaria	686	4

2. Different plant organs accumulate different quantities of zinc. Generally, roots and leaves take up most zinc, and stems and inflorescences least, but in some species the flower heads seem to accumulate more. The pattern of distribution depends both on the species and on the metal concentration (Ernst, 1965b; Nicolls *et al.*, 1965; Cole *et al.*, 1968; Howard-Williams, 1969).

3. The quantity of zinc in plants changes with the growing season and often shows an increase throughout the season (Ernst, 1965b; Howard-Williams, 1969).

4. The quantity of zinc in plants is related to the amount of zinc in the soil often in a clearly linear pattern (Nicolls *et al.*, 1965). Occasionally the plants seem to take up more zinc than is present either in "available" form (e.g. *Thlaspi alpestre*, Ernst, 1965a) or in total amount (e.g. various lichens, Maquinay *et al.*, 1961; Lambinon *et al.*, 1964).

5. The cellular resistance of plants to solutions of different zinc concentrations is related to the amount of zinc they absorb (Baumeister, 1967).

These findings are confirmed by studies of zinc uptake by tolerant plants in water culture. Baumeister (1967) found a fairly linear relation between the amount of zinc in *Thlaspi alpestre* and *Armeria maritima* and the amount in the culture solutions even though the absolute amounts in the two species differed enormously. Turner and Gregory (1967) found increasing uptake with increasing concentration of zinc by *Agrostis tenuis* (tolerant and non-tolerant) in water culture. The concentrations (ppm dry weight) in the plants were in excess of the concentration (ppm in solution) in the solutions after only seven days at all concentrations. The roots showed far greater accumulation than the shoots, particularly at higher concentrations. This suggests that the values found by workers for shoots of plants growing in the field may underestimate the overall uptake of zinc by plants. Ernst (1968c) looked briefly at the zinc content of roots as well as shoots of plants in the field. *Thlaspi alpestre* showed greater accumulation in shoots than roots, whereas *Minuartia verna* showed similar root and shoot contents. There therefore appears to be variation between species. However he also showed (Ernst, 1968d) that under experimental conditions uptake was affected by phosphate: at high phosphate concentrations zinc was much higher in roots than in shoots. Zinc uptake clearly requires further study.

Zinc therefore is readily taken in by plants growing on zinc-contaminated soil, and nowhere in the literature is there any evidence of these plants having an exclusion mechanism to enable them to survive on contaminated soils. The tolerance mechanism of zinc must be internal.

2. *Copper*

Studies on copper uptake have been fewer and less extensive than those on zinc uptake.

The earliest reference is probably that of Bateman and Wells (1917), who found appreciable quantities of copper in plants (both living and dead) on copper-contaminated soils. Figures of 2000–6000 ppm were obtained for *Plantago*, *Agropyron* and *Dasiophora*, while lower values were found in *Medicago*, *Equisetum* and *Trifolium*. Dead vegetation on the whole contained more copper than living plants. Prat and Komarek (1934) found that plants of *Agrostis alba* (*stolonifera*) and *Melandrium sylvestre* growing on soils rich in copper (1–39% Cu) contained 0·2–3·25% copper in the ash. Persson (1956) again found that "copper mosses" do in fact take up this element in appreciable quantities.

The only detailed investigation has been on plants on metal outcrops in Australia (Nicolls *et al.*, 1965) and it has revealed a pattern common to several species. Copper uptake in the above ground parts stayed low and constant at low levels of soil copper, but at certain higher soil copper levels this "resistance" to uptake seemed to break down. Above this level the quantity in the plant tops increased abruptly and at only slightly higher levels in the soil no plants were found. Species differed both in the overall copper content and in the level at which the sudden increase in copper uptake was seen. These differing reactions to copper in the soil suggest that the copper did not just become available to the plant at a given total soil level, but that a genuine exclusion mechanism was in action at low copper levels. This is supported by the finding that values of copper even a little above the level at which there is greatly increasing uptake are lethal. Moreover, the species found in the most toxic areas, *Polycarpaea glabra*, takes up very little copper (a maximum of 20 ppm in the leaves) on soils containing 10 000 ppm. If the results quoted by Ernst (1965a) are re-examined in the light of this work of Nicolls *et al.* (1965), a remarkably similar pattern emerges, with both *Silene inflata* and *Minuartia verna* showing constant low uptake at low levels but a sudden increase to high uptake at high level. Although the results are based on far fewer samples than those of Nicolls *et al.*, the similarity is startling.

There is therefore evidence that the mechanism of copper uptake is different from the uptake of zinc. These results are supported by other less extensive studies. Vogt *et al.* (1943) and Vogt and Bugge (1943) analysed plants growing over an exposure of copper and showed that they did not differ in copper content from those growing off the ore. However, there were marked differences in zinc content.

Duvigneaud and Denaeyer-de Smet (1963), who fully review the literature on copper content of plants on normal soils, studied a wide range of species on different soils. They found that most species growing on soils containing often 1000 ppm copper, only showed slightly enhanced copper contents (two to three times that of plants on normal soil in the same area). Only three out of twenty-four species accumulated more than 1000 ppm dry weight. Ernst (1968b) found appreciable quantities of copper in *Minuartia verna*, but low values in four other species. Roots consistently contained greater amounts of copper than shoots: shoot values may therefore underestimate the copper content of plants. As with zinc, the relative distribution of copper between root and shoot would repay further study.

Lange and Ziegler (1963) again found variations between species of lichens in copper content, those on copper soils generally containing slightly more copper.

These results are further supported by investigations on copper and zinc in plants growing on normal soil. McHargue and Roy (1932) found that there was little variation in copper content of tree leaves over the growing season, but that zinc showed considerable variations. Holmes (1964) noted that whereas copper in plants rarely varies more than 5–15 ppm the zinc content can vary from 20 ppm to 10 200 ppm.

Despite the general consistency of patterns of uptake of copper, the recurring evidence that species may differ cannot be dismissed. Jacobsen (1967) provides perhaps the best evidence for the plants occurring on a single natural contaminated area in Rhodesia. Following the method of Thyssen (1942), he determines an "enrichment factor" which is effectively the copper in twigs as a percentage of that in the soil. This gives values of up to 5% for some species such as *Combretum zeyheri* and only 0·5% for other species such as *Brachystegia spiciformis*. The species of the first group tend to be capable of growing on high copper soils. In all species, however, the amount of metal in the plant again seems to remain remarkably constant over a wide range of soil copper values.

There has only been one study of copper uptake from culture solution, namely that of Bradshaw *et al.* (1965). Copper-tolerant and non-tolerant *Agrostis tenuis* showed a slight marginally significant increase in copper content of the shoots (leaves) with increasing copper concentration in solution. However, the roots showed a large highly significant increase.

The roots absorbed far more copper (maximum 3253 ppm dry weight) than the shoots (maximum 41 ppm dry weight). The level of copper concentrations used was low (0–1·5 ppm), showing a remarkable accumulation of copper by *A. tenuis*. These results emphasize the importance of studying both root and shoot in uptake studies.

In the absence of extensive and detailed studies of uptake from solutions, it is difficult to interpret studies relating soil copper content to plant copper content. The situation is further complicated since high copper contents may go hand in hand with low copper availability (see Sections IIc4 and VIIA). Nevertheless, some tentative conclusions can be drawn. Firstly, there are large species differences in pattern of copper uptake. Secondly, copper uptake occurs and tolerance in some species must be due to an internal mechanism. Thirdly, in other species an exclusion mechanism is probably also involved, but its precise nature needs re-examination. The mechanism may well not be a true exclusion mechanism, but a complexing mechanism in the plant roots which prevents copper from reaching the upper parts of the plant until the root system is swamped. Fourthly, the pattern of copper uptake seems very different from that of zinc uptake.

3. Lead

Studies on the uptake of lead have been very few. Jensch (1894) showed that whereas *Tussilago farfara* and *Polygonum aviculare* on contaminated soils contained zinc, no lead was detectable, even though the soil contained 0·72–1·06% of this element. These results are similar to those found by Nobbe, Bressler and Will (quoted in Schwickerath, 1931, but no reference given): if equal quantities of lead and zinc are given to a plant, then a smaller quantity of lead than zinc is taken up.

Nicolls *et al.* (1965) also dealt with lead uptake. The pattern here resembles that of copper rather than zinc, in that the uptake is constant with increasing levels of soil lead, till a certain point is reached when uptake becomes unrestricted, and rises abruptly. The species also are rarely present when the soil lead value is above the level at which there is a sudden increase in lead uptake.

The pattern of lead uptake by plants has, therefore, been rarely studied, and little can be concluded about the mechanism involved, except that it seems to be similar to copper.

4. Nickel

There has been little systematic work on nickel uptake. However, preliminary work by Wild (1970) shows that for many species the pattern of uptake appears to resemble that of copper: levels in the aerial parts of the plant remain low (around 100 ppm) irrespective of external concentration. However, this was not true for two species, *Vellozia equisetoides* and *Dicoma macrocephala*, where the uptake reached 1000 ppm. Such findings raise very interesting problems and suggest that mechanisms of tolerance to nickel may differ in different species.

C. INTERNAL MECHANISMS

From the previous sections it is clear that higher plants on contaminated soils frequently take up toxic levels of metal ions and that the tolerance mechanism must at least in part be internal. This is almost certainly true for other organisms in toxic areas or laboratory races selected for tolerance. As might be expected, the mechanisms whereby such tolerance is achieved are diverse. But only two groups of organisms, the yeasts (especially *Saccharomyces cerevisiae*) and higher plants (especially *Agrostis tenuis*), have been studied in any detail in this respect and they will be considered first.

1. Yeasts

The mechanisms whereby laboratory-selected strains of yeast achieve tolerance to a wide variety of metals has been reviewed by Ashida (1965) and more briefly by Turner (1969). The following section will therefore summarize the work rather than attempt an exhaustive review. Copper-tolerant strains of yeast *Saccharomyces cerevisiae* appear to differ from non-tolerant strains most obviously in their sulphur metabolism. Tolerant strains appear brown in colour when cultured with copper as a result of the production of a large amount of copper sulphide. Hydrogen sulphide appears to be released as a by-product of an overactive cysteine pathway; it combines with copper and the sulphide is deposited in and around the cell wall.

This however does not appear to be the only mechanism of copper resistance in these strains: strains resistant to very high copper concentrations produce less hydrogen sulphide, respiration-deficient strains produce no hydrogen sulphide and yet are still copper resistant, and genetic recombinants can be obtained which have high copper resistance but low hydrogen sulphide production. Other mechanisms may include the presence in the cell of copper-binding substances such as cysteine or thiol-based compounds. There is also evidence of a cuproporphyrin complex in tolerant strains. Metabolic differences include a greater dependence on aerobic respiration in the resistant strain (since the fermentation pathway is more sensitive to metal ions) and a greater dependence on NADP as opposed to NAD-linked reactions in the tricarboxylic acid cycle. There was evidence of changes in the RNA of copper-resistant strains of a related species, *S. ellipsoideus*.

Tolerance to other metals (Cd, Co, Ni, Ag) do not appear to depend on changes in sulphur metabolism. They do not show clear-cut cross-tolerances with each other or with copper, suggesting that different mechanisms are almost certainly involved in the different strains.

2. Higher plants

The mechanism of metal tolerance in higher plants from mine soils has been reviewed and studied with particular reference to zinc tolerance in *Agrostis tenuis* by Turner and Gregory (1967), Turner (1969, 1970) and Peterson (1969). Turner (1967) showed that sulphur deficiency appeared not to affect the tolerance of zinc- and copper-resistant strains differentially from non-tolerant strains, suggesting that changes in sulphur metabolism are not important in the tolerance mechanism in higher plants. Subsequent studies involving cell fractionation and differential centrifugation (Turner and Gregory, 1967; Turner, 1969, 1970; Turner and Marshall, 1971) showed that in tolerant strains zinc and copper are concentrated in the cell walls, and thereby prevented from entering more sensitive sites in cell metabolism. The level of zinc absorbed by the cell wall was correlated with zinc tolerance of different plants. Further attempts at localization showed that zinc was not released by proteases, suggesting localization in a non-protein, possibly carbohydrate, fraction although Woolhouse (1970) disagrees with this.

Alteration in the external level of one heavy metal did not influence the distribution of the other metal, suggesting the mechanisms are specific for each metal (cf. Gregory and Bradshaw (1965), who failed to find cross-tolerance for copper and zinc). These results were confirmed by Peterson (1969) using chemical fractionation of the cells of zinc-tolerant *Agrostis tenuis* and *A. stolonifera*. He found that the pectate fraction of cell walls of tolerant plants contained five to six times as much zinc as in non-tolerant plants. The distribution of zinc in copper-tolerant plants was similar to its distribution in non-tolerant plants, confirming that the tolerance mechanism for the two metals is different. However, *Agrostis stolonifera* differed from *A. tenuis* by absorbing an appreciable quantity of zinc on the soluble RNA fraction, suggesting certain species differences as regards zinc distribution.

Various points are still not clear. It is not clear to what extent the cell wall is important in metal tolerance in organisms other than *Agrostis* spp. Evidence supporting the general importance of the cell wall in metal ion inactivation comes from the demonstration that plants from normal soils often contain large quantities of metals in the cell walls, and from work on tolerant micro-organisms (see Turner (1969) for review and references). There is also evidence from less rigorous work on metal-tolerant plants.

Ernst (1968c) showed that the zinc content of expressed sap from *Thlaspi alpestre* and *Minuartia verna* increased only slightly with large increases in total zinc concentration in the plant. There was more zinc in the sap of *Thlaspi alpestre* than *Minuartia verna*, suggesting species differences. Subsequently he showed (Ernst, 1969a) that the zinc concentration of cell sap from various organs of *Silene vulgaris* increased during the growing period, suggesting increased saturation of a complexing system in the cell walls. However, he also noted that a large proportion of the zinc was water extractable and probably found in the vacuole (e.g. 74% in *Cardaminopsis halleri*), and suggested that the vacuole was an important site of metal ion storage. Throughout his experiments Ernst did not, however, use non-tolerant material for comparison.

Gambi (1967) showed histochemically that nickel was concentrated in the epidermis and sclerenchymatous areas (cells with no living contents) between vascular bundles in *Alyssum bertoloni*, a nickel accumulator found on serpentine soils.

A cell wall complexing system suggests a mechanism with a finite capacity for absorbing metal. It is interesting that an apparent threshold nature of tolerance to increasing metal levels is observed in artificial selection for tolerance (see Section VA3) suggesting a "mop-up" system capable of being saturated.

There is evidence that tolerant plants are metabolically different from normal plants at low metal levels but it is difficult to know whether such differences are related to the tolerance mechanism or the result of independent selection for physiological adaptation to other aspects of the mine habitats (see Section VF). And at high metal levels other metabolic differences become apparent between tolerant and non-tolerant *Agrostis tenuis* (Gregory, 1965; Turner, 1967) but again it is not clear whether these are the result of differential sensitivity to metal ions or whether they are related to the mechanism of tolerance. Repp (1963) and Gries (1966) have both reported increased general protoplasmic resistance of tolerant strains and this is evidence that systems other than those in the cell walls may contribute to tolerance to heavy metals. However, the technique for testing protoplasmic resistance does not critically eliminate the possibility that the cell wall may be important.

Since differences in enzyme properties of tolerant strains of microorganisms have been reported, it is clearly worth while looking for similar differences in higher plant enzymes. The effect of lead ion concentration on the activity of certain root cell wall enzymes of *Agrostis tenuis* has been investigated by Woolhouse (1970). The rate of hydrolysis of *p*-nitrophenyl-phosphate by root tips of normal non-tolerant material was affected by concentrations of lead as low as

1 mM, whereas that of lead tolerant material was not affected until the concentration reached 10 mM. It is argued from this that there must be different forms of the phosphatic enzymes concerned. While this is possible, an alternative explanation is that the enzymes are the same but that in the tolerant material the heavy metal complexing mechanism in the cell wall is preventing the metal reaching the cell wall enzymes in the same way that it appears to protect other sensitive sites e.g. on mitochondria. Further investigation will be very rewarding.

A remarkable feature of metal tolerance is its specificity. Tolerance to one metal does not generally confer tolerance to other metals (see Gregory and Bradshaw, 1965; Turner, 1969, for discussion and review). The only observed instances of observed cross-tolerances are between zinc and nickel in *Agrostis tenuis* (Gregory and Bradshaw, 1965), and copper and chromium in desmids and higher plants from normal soils and copper-tolerant mosses (Url, 1955, 1956). Although many explanations of such specificity can be postulated, its precise nature remains unexplained.

3. *Other organisms*

The mechanism of tolerance in both yeasts and higher plants involves in part at least removal of toxic ions from general metabolism. Analogous mechanisms are known in other organisms. Rabanus (1931) and Shimazono (1951) both suggested that the metal tolerance of brown-rot fungi was the result of the copper ions being combined to produce insoluble oxalates. However, the earlier work of Richards (1925) demonstrated that copper tolerance did not confer zinc tolerance, yet a non-specific metal-complexing mechanism such as oxalate production would be expected to do so. An elegant example of a metal-complexing mechanism responsible for heavy metal tolerance has been shown for *Aspergillus niger* (Ashworth and Amin, 1964). Mercury tolerance in *A. niger* is due to a pool of non-protein sulphydryl groups that protects enzyme systems by forming complexes with the mercury as it enters the thallus. *Rhizoctinia solani* and *Pythium ultimum* do not have this pool of non-protein sulphydryl groups and are thus susceptible to mercury poisoning. These non-protein sulphydryl groups can combat zinc, nickel, lead, copper and silver poisoning (Ashworth, 1965).

Tonomura, Maeda and Futai (1968), Tonomura, Maeda, Futai, Nakagami and Yamada (1968) and Tonomura, Nakagami, Futai and Maeda (1968) have shown that mercury-resistant bacteria bind the components probably at the cell wall and convert them to a more volatile, but as yet unidentified, form so that mercury is lost from the medium. Rapid growth resumes as the mercury concentration of the medium falls.

Differential uptake may also at times be important: Earley (1943) found that zinc-tolerant soybeans did not take up as much zinc as did susceptible varieties. However, it is possible for tolerant organisms to have a low metal content when cultured at high metal levels which is not due to an active exclusion mechanism: an apparent lack of permeability may be the result of metal toxicity and injury which disrupts cell membranes. Terui *et al.* (1960) found less arsenic in tolerant hyphae of *Aspergillus oryzae* as the arsenite level was increased; however, increasing arsenite content of the medium was paralleled by increasing hyphal damage. Yamamoto (1963) found that in strontium-resistant yeast cells, the protein synthesis of the resistant cell is aided by the morphological stability of the ribosomal particles under conditions of hypertoxicity produced by excess strontium. Repp (1963) found that the protoplasm of *Silene vulgaris* was generally highly "shock resistant" when exposed to heavy metals. She also found that metal-tolerant races of *Tussilago* and *Taraxacum* had protoplasmic changes that allowed them to overcome the "shock" of excess heavy metal ions.

Increased concentrations of a metabolite may antagonize the effects of the metal. This mechanism is analogous to increases in a metal-binding agent as with sulphur in resistant yeast cells (Ashida and Nakamura, 1959). Selenium and sulphur have been shown to have competitive effects in metabolism: increases in sulphur nutrition antagonize selenium poisoning by competing for sites of reaction rather than by binding with the toxic metals (Rosenfeld and Beath, 1965). Lohrmann (1940) found magnesium interfered with boron and mercury poisoning in several species of fungi.

Wood and Sibley (1950) showed that dead tissues of oat plants always had higher concentrations of zinc than did normal photosynthesizing tissue, suggesting the possibility of "excretory" mechanisms being important in metal tolerance. However, there is no evidence for this in tolerant plants.

D. CONCLUSION

The mechanisms whereby plants combat toxic levels of heavy metals are very varied, and in yeasts in particular there is evidence that the whole metabolism is altered in copper-tolerant strains. In higher plants the tolerance mechanism appears to be designed to keep metal ions away from the active sites of metabolism by chelation in the cell wall. It is interesting that, in *Agrostis* at least, similar though independent complexing systems are present in the cell wall for both zinc and copper. This and the circumstantial evidence from other organisms argues that the cell wall may have a general function, hitherto unrecognized, in removing excess levels of toxic ions even in relatively normal situations.

A problem that remains to be solved is the reason for the high degree of specificity of metal tolerance. Even copper-tolerant yeasts which produce the relatively unspecific inactivating agent, hydrogen sulphide, do not show cross-tolerance to other metals. In higher plants the specific nature of the mechanism has to be reconciled with the fact that tolerance appears to be determined not by a single gene but polygenically (see Section VI). Genetic studies, particularly if combined with biochemical investigations, could help throw light on the tolerance mechanism and its specificity.

The various mechanisms of tolerance all suggest that special systems are developed by organisms to cope with excess metal ions: tolerance appears not to be achieved by producing excessive amounts of substrates (e.g. enzymes) that have been previously known to complex with metals in biological systems. This is understandable in the case of metals such as lead and mercury which are not normal components of metabolism, but more surprising with trace elements such as copper and zinc. However, the requirement for a special tolerance apparatus may explain why only a few organisms appear to be capable of evolving tolerance to any degree, and why none have been shown to possess an inherent metal tolerance (see Section Vb). Nevertheless, the cell wall may have some general importance as regards metal binding.

The gross metabolic alterations that occur in tolerant races may in part explain their poorer competitive ability and changes that have been noticed in characters other than tolerance. However, the situation here needs clarifying since independent selection for other characters will explain both phenomena. Investigating the tolerance mechanism in artificially selected material and in segregating progenies should help to clarify these problems.

The nature of the tolerance mechanism, as well as being of intrinsic physiological and biochemical interest, clearly has considerable relevance to the evolutionary and ecological aspects of metal-tolerant plants. It is encouraging that an integrated, consistent picture is beginning to emerge, and further work should prove extremely rewarding.

VIII. Practical Applications of Studies of Metal-tolerant Organisms

A. Biogeochemical Prospecting

The role of plants as indicators of soils contaminated with heavy metals has already been discussed (Section IIa). The use of these plants to locate natural outcrops of metal ore is less well documented. They are frequently mentioned, but this is more often in terms of their curiosity value. Over the past twenty-five years the use of plants in prospecting has developed in more sophisticated ways. Valuable reviews are provided by Cannon (1960a), Malyuga (1964) and Cole (1965) so that only the general perspective will be given.

1. Vegetation changes

Many of the earlier prospectors relied on the look of the land and the plants growing there to tell them whether there was a chance of a strike. Old textbooks on mining recommended that attention should be paid to plants (Foster, 1894).

The value of plants as indicators can be either general or particular. General changes in species occurrence is an indication of some sort of soil anomaly, and provides an initial sign that there is a soil and underlying rock change (Malyuga, 1964). There is no doubt that this technique has been widely used by prospectors who did not pay attention to individual species.

Changes in particular species can be more valuable, for these can indicate the presence of a particular metal. But whether or not they do is another matter as has already been seen in Section IIA1. The problem is that many of the species supposed to be associated with a particular metal are in fact associated with a particular rock or soil type. This is very clear in the case of serpentine which has a very characteristic depauperate appearance and flora associated with high magnesium level (Walker, 1954); it is a rock in which chromium and nickel is found. As a result of this chromium ores in serpentine were discovered in Maryland in 1818 (Singewald, 1928). But there are many other serpentine areas with an equally characteristic vegetation without metals. Another example is *Becium* (= *Ocimum*) *homblei* which has been reported to grow only on soil containing more than 100 ppm copper in Rhodesia (Anon., 1959). But it has recently been shown in an intensive investigation to grow widely on non-copper soil, on granite sandveld, being favoured by very poor soils and lack of competition from *Becium obovatum* (Howard-Williams, 1970). *Vellozia equisetoides* is similar to *B. homblei*; it is common on the skeletal soils of quartzite slopes yet is widespread and characteristic on copper soils (Wild, 1968).

If a species has a reasonable association with a particular mineral-bearing rock it can be a valuable indicator even if that association, as for *Vellozia equisetoides*, is not absolute and is due to other soil characteristics and absence of competition. The indicator plants that have, for instance, been most valuable for copper are *Elshotzia haichowensis* in China (Tsung-Shan, 1957), *Acrocephalus robertii* in Katanga (Duvigneaud, 1958), and *Becium homblei* in Rhodesia (Anon., 1959), all belonging to the Labiatae. In Australia in the nineteenth century, the pink *Polycarpea spirostylis* was recognized as a copper indicator by prospectors in Queensland (Cole, 1965). Areas of copper in Sweden have been located merely by examining the localities from which herbarium specimens of "copper mosses" had been collected (Persson, 1948). Uranium ores in the Colorado Plateau and in Peru have been detected by the occurrence of members of the genus *Astragalus* (Cannon, 1957, 1960b). But there are many other records in the literature: these are listed in Table Ib, and are discussed by Cannon (1960a) and Malyuga (1964). Indicator species pose very interesting problems: it is a pity no critical ecological and genetical investigations have been made of any except *Becium homblei* (Howard-Williams, 1970).

It is often better when reliance is not placed on the occurrence of a single species but on a group of commonly associated species—a lode assemblage. This is clearly indicated by the very extensive work recently carried out in Australia (Cole, 1965; Nicolls *et al.*, 1965; Cole *et al.*, 1968), and is supported by work in the U.S.A. (Cannon, 1960a), Russia (Malyuga, 1964; Chikishev, 1965) and Rhodesia (Wild, 1968, 1970). *Polycarpeae glabra*, *Eriachne mucronata*, *Bulbostylis barbata*, *Fimbristylis* sp. nov. and *Tephrosia* sp. nov. are all associated with lodes in the Dugald river area of Australia containing zinc, copper and lead, but they do not always all occur together. Their occurrence has revealed new areas for intensive prospecting. These species (particularly *Tephrosia* sp. nov.), and a few others, were found to be associated with similar mineralization in other areas of North Australia (Cole, 1965). But they may sometimes be found in areas where metals are not present but the soil is anomalous in other respects, and they may be absent on soils where the metal content is extremely high (Cole *et al.*, 1968). They may also be replaced by other species if the surrounding bedrock changes markedly (Nicolls *et al.*, 1965). As a result of the influence of climate and soil factors on the occurrence of indicator species, Wild (1968) advocates the need to become familiar with the assemblage of species on one copper outcrop in order even to recognize further outcrops in the immediate vicinity. This applies also to nickel (Wild, 1970).

Areas where the concentration of metal is very high may be completely bare of vegetation, or particular sensitive species may be absent. Bareness or lack of species can, therefore, be an important indicator of metal occurrence. This technique has been used in prospecting for copper in Armenia and Rhodesia. In the Congo high copper areas are recognized by the absence of trees in a predominantly forested area (Cannon, 1960a). Sometimes an indication of actual levels of metal in the soil is obtained by the appearance of particular plants. High zinc causes chlorosis in many species. Copper gradients in Katanga are recognized by the stunting of *Protea goetzeana* (Duvigneaud, 1958).

2. The metal content of plant parts

The value of plant analysis in prospecting stems from two characteristics of plants growing over contaminated soil. Firstly, as has been reviewed in Section VIIB, plants take up appreciable quantities of metal into their aerial parts. This is true of plants growing over non-toxic (but metal-containing) soils as well as toxic soils. Secondly, the roots of trees in particular penetrate to the lower depths of the soil, often to the bedrock. Sampling aerial parts of a plant may therefore supply information on metal content of the bedrock without recourse to drilling operations. But the disadvantage of the method in comparison with more normal geochemical techniques is the need for ashing the plant material, and lack of perfect correlation between plant and soil metal contents.

The method of prospecting has been examined, initially independently, in Canada, Scandinavia and Russia, and recently in Australia. It has been reviewed by Hawkes (1948, 1957).

In Canada, Warren and Howatson (1947) and Warren and Delavault (1948) made pioneer studies on copper and zinc contents of trees lying over known ore bodies and trees on normal soil. They concluded that newly grown twigs and leaves gave the most reliable results, and produced empirical data which, when exceeded, suggested abnormally high soil metal concentration. Absolute values and Zn/Cu ratios were both used. Even at this stage differences in uptake according to soil type were noted: considerably less was taken up on calcareous soils (see Section VIIB). Warren, Delavault and Irish (1949) extended these results and modified the empirical values. In the process of this work they discovered an unknown vein lying 15 ft below the surface. Warren and Delavault (1949) quoted a Cu/Zn ratio of 0·10–0·15 as about the normal range, and a value less or more than this as being indicative of zinc or copper anomalies. Warren and Delavault (1950a) showed that the technique could be applicable to the detection of gold. White (1950) made a more systematic study of metal contents of plants over ore deposits using mapping techniques. In this way he was able to detect the existence of metal shadows around ore deposits. Shallow deposits produced a small abrupt shadow whereas deeper deposits led to a shadow as much as fifty times as wide as the ore deposit itself. He concluded that the metal shadow can be detected with an overburden in excess of 30 ft. This is confirmed by Clarke (1953). The sampling technique has been further sophisticated by Warren, Delavault and Fortescue (1955).

In Norway Vogt (1939, 1942a, 1942b), Vogt and Braadlie (1942),

Vogt and Bugge (1943), and Vogt *et al.* (1943) made a similar study in the Røros area of Norway, but beyond showing the higher level of metals, particularly copper and zinc, in plants from contaminated areas, they did not champion the value of this technique in prospecting. Rankama (1947) briefly reviewed some results of this type of study in Europe and quotes examples where it had been successful. Subsequently, he showed that nickel could only be detected if inorganic material covering nickel deposits was only a few metres thick (Rankama, 1954).

The Russian work on biogeochemical prospecting has been fully reviewed by Malyuga (1947, 1964). He considers some of the factors that affect the form of the metal shadow, (e.g. depth of ore body, aspect, root depth) and gives extensive details of examples where plant analyses have been of practical value, in outlining ore bodies and in making new discoveries. It is not possible to describe the wealth of information given by Malyuga. But it is clear that analyses of a wide range of plant material have been valuable in the U.S.S.R. in the detection of ore bodies of a wide variety of metals and can only cost one-tenth as much as normal exploration techniques involving drilling, trenching, etc. But care is necessary in the sampling processes because of the variation between species, between different parts of the plants, between seasons, and between different climatic regions.

The fact that the technique is not perfect and the correlation between soil and plant metal content can go astray was pointed out by Marmo (1953), who showed that Cu levels in soil are reflected by Cu levels in plants at low levels of soil copper but not when soil copper levels are high.

Recently, extensive analyses have been carried out on a wide range of Australian species (Nicolls *et al.*, 1965). The relationship, in all species analysed, for both copper and lead is one in which the plant metal content does not increase with increasing concentration of metal in the soil until a value of about 2000 ppm, above which there are only few records because the plants are eliminated. The same relationship for copper is described by Malyuga (1964). For zinc, however, the relationship is linear. These results suggest that the method may not be very useful for copper and lead although it is very satisfactory for zinc. But the evidence is confusing since copper levels in vegetation have been correlated with the presence of known and previously unknown ore bodies in the Gaspé peninsula (Riddell, 1952) and in British Columbia (Warren and Delavault, 1950b).

Cobalt, chromium, copper and nickel uptake by a New Zealand serpentine flora has been critically examined by Lyon *et al.* (1968). In the six species examined there was a reasonable correlation between metal levels in soil and plant for all the metals except copper, in *Cassinia vauvilliersii*, *Leptospermum scoparium* and *Hebe odora*. For the other species correlations for any metal were weak or absent. From this *Cassinia vauvilliersii* was suggested as a good species for biogeochemical prospecting, but it would not appear from the data to be useful for copper.

An extensive list of those cases where prospecting by plant analysis has been undertaken, and their success, is given by Cannon (1960a). The absorption of metals by plants is a complex phenomenon and not clearly understood (see Section VIIB). Part of the complexity is due to complexities of soil and climate, but part is due to the complexities of species. We are perhaps prepared to find that species differ in their patterns of uptake. But no one using plant analysis has considered the problem that the species being analysed may well consist of completely different populations with contrasting patterns of uptake.

3. *The metal content of the humus layer*

Goldschmidt (1954) showed that rare elements are 4–200 times more concentrated in coal than in the average rock. He concluded that humus layers tended to accumulate elements which were brought to the surface by deeply penetrating roots. This was independently realized by Vernadskii (see Malyuga, 1964 for full discussion) and has since been used in the Soviet Union in combination with plant analyses and detailed mapping techniques. It is an extension of the geochemical techniques which are very widely used. It appears to be of most value for copper which forms stable complexes with organic material.

4. *An integrated approach*

Recent studies (Malyuga, 1964; Cole, 1965; Nicolls *et al.*, 1965) have pointed the way towards a large-scale exploitation of these techniques. The first stage of such prospecting involves recognition of a suitable area for further study: this will be decided by geological studies and by associated general changes in vegetation. The next stage is a systematic study of the area: this may be by grid sampling or transect sampling, covering the whole or certain parts of the area. A survey is made of the vegetation and of individual species: and plants, humus and soil are analysed. In this way it is possible to locate metal shadows and specify the position and composition of ore bodies. More generally it can indicate promising areas for drilling operations.

This integrated method is almost certainly the most successful. Its value lies in the fact that useful information can be obtained without drilling and using relatively simple sampling techniques and simple analyses. It certainly provides information of considerable interest to the botanist: it is only a pity that parallel information is not being obtained on the population structure and physiology of the important species.

B. ORE EXTRACTION

Bryner *et al.* (1954) showed that micro-organisms isolated from leaching effluent promoted the solubilization of iron and copper sulphide ores by a factor of 10 to 20 over the rate of solubilization in sterile culture. Subsequent work confirmed these results and identified the active bacteria as *Thiobacillus ferro-oxidans*, *T. thio-oxidans* and *Ferrobacillus ferro-oxidans* (Beck and Elsden, 1958; Bryner and Jameson, 1958). Bryner and Anderson (1957) demonstrated that micro-organisms were also involved in the solubilization of sulphide ores of nickel, cobalt and molybdenum.

Recently work in this field had been reviewed by Razzell and Trussell (1963), Beck (1967) and Ehrlich and Fox (1967). The technique of extracting copper from low-grade ores involves three stages: the leaching process itself, precipitation of the copper by addition of iron, and oxidation of the ferrous sulphate to ferric resulting in precipitation of the iron with the recovery of water for the leaching process. Beck (1967) studied the role of *Thiobacillus ferro-oxidans* in each of these stages. He showed there was a decrease of bacteria in the precipitation plant, a large increase in the oxidation pond, and a slight decrease during passage through the ore dump. The latter surprising fact led Beck to speculate that large numbers of bacteria were filtered out of solution and that the leaching was as much a result of increased temperatures and (possibly triggered off by bacterial activity) chemical oxidation as biological oxidation. The role of bacteria in leaching operations has been further examined by Ehrlich and Fox (1967), who showed that bacteria did not just oxidize the sulphide ores, but helped extraction indirectly by releasing the ferric ion which then oxidized copper sulphide non-biologically (ferrous iron), and by depressing iron release from the ore in some unknown fashion. Ehrlich and Fox (1967) demonstrate that a wide variety of factors may affect ore extraction, including the interaction of bacteria with a wide range of micro-organisms (Ehrlich, 1963b).

It is clear from these studies that the role of bacteria and other micro-organisms in leach operations needs further definition particularly if any attempt is to be made to optimize the conditions for such leaching operations. The bacteria are likely to be present in most mine dumps and optimizing conditions for leaching may permit commercial exploitation and/or subsequent recolonization by higher plants of tips depleted of their metal content.

C. THE ESTABLISHMENT OF VEGETATION ON TOXIC WASTE MATERIALS

The toxic materials left by mining operations are unattractive to the eye and can be a considerable source of pollution of surrounding areas. The piles of country rock produced to get at the ore bodies are coarse, very low in major nutrients, and may contain small quantities of toxic metals. The tailings, produced by the dressing procedures which concentrate the ore, because of the economics of extraction procedures often contain as much as 1% of metal, although modern techniques may reduce this five-fold. The tailings contain not only the remains of the metal that has been extracted but also quantities of the other metals that were present in the ore and not extracted. These metals are released slowly as soluble compounds as the ore weathers, and may continue to be released for centuries.

As a result the waste heaps have very little plant growth upon them, owing to the combined effects of toxicity, lack of nutrients and physical conditions. The particle size of the tailings is often sufficiently small for it to blow in dry weather on to surrounding areas. Particles are nearly always fine enough to be able to be carried by run-off during storms into water courses, often for long distances, unless great care is taken in the establishment of dams and settling ponds. There will always be a certain amount of leaching of soluble salts from waste heaps, which will be enhanced if the heaps are unstable. Thus the pollution of streams by dissolved heavy metals in the neighbourhood of mine workings can be considerable: levels of 500 ppm can easily be found.

The establishment of a vegetation cover on toxic waste heaps will do a great deal to minimize pollution, particularly where the pollution is due to the movement of solid particles, and will also improve the appearance of the heaps. But the conditions of toxicity and poverty of nutrients, and the physical nature of the material make normal techniques of vegetation establishment impossible (Hooper and Newton, 1935).

The most obvious technique that will be effective is to cover the toxic material with a layer of normal soil and establish a herbage cover or trees in the normal manner. This is, however, exceedingly expensive, usually at least six times the normal for non-toxic areas.

As a result experiments have been carried out in several places to find ways in which the toxicity, nutrient poverty and physical conditions of the material can be ameliorated more simply by addition of less expensive materials. Physical and chemical methods are possible (Dean *et al.*, 1968) but the most promising additives are organic wastes such as sewage sludge and town refuse. With these, swards of normal herbage species have been established over highly toxic smelter waste in the lower Swansea Valley, so long as the ameliorating material is more than about ten centimetres thick (Weston *et al.*, 1965; Street and Goodman, 1967). Sewage sludge is particularly successful since its high organic matter content effectively ameliorates all three conditions. But the effect of these materials may be short term: when the organic matter decays the toxicity will return, and the vegetation will eventually suffer and die.

An alternative approach is to take advantage of the natural evolutionary processes that have already occurred and employ species and populations which are already tolerant to the toxicity present, together with fertilizer to ameliorate nutrient lack. Preliminary experiments with smelter waste in the lower Swansea Valley using vegetative material of various grass species showed that the performance of tolerant material was superior to that of non-tolerant (Gadgil, 1969). But, not unexpectedly, this superiority was not found when domestic refuse or sewage sludge was used, when all populations grew well.

This method has now been investigated using seed on a variety of mine tailings in Wales and Northern England, with the addition of different amounts of fertilizer (Smith and Bradshaw, 1970). The superiority of tolerant material was remarkable: non-tolerant material was almost dead in nine months while tolerant material grew almost as well as if on normal soil. The effect of fertilizer was very marked indeed, slow release forms giving best growth.

In both these experiments it appears that physical features of mine waste are not severe enough to prevent growth. However, in drier climates they may be important, but perhaps can be overcome by sowing in the wet season or by using mulching techniques (reviewed by Peterson and Monk, 1967).

The technique of using metal-tolerant populations has not been explored in a wide enough variety of sites. But initial results are promising and fit in with expectations. There is such a wide variety of species in all parts of the world in which tolerance is known or can be presumed to occur, that it seems a method capable of widespread use. Many of the species are vegetatively vigorous and perennial, and some are leguminous and therefore will perhaps be self-sufficient for nitrogen unless N fixation is precluded by toxic soils. The method should be investigated further.

IX. Conclusion

The plants growing in habitats contaminated with toxic levels of heavy metals have attracted interest for over a century. Since metal-contaminated areas are only a minor component of the environment, the subject of metal tolerance has on the whole been regarded as an area of marginal relevance to most other fields of investigation. The interest that has arisen has come from ecologists, evolutionary geneticists, physiologists, and applied biologists particularly in relation to prospecting for metal ores. The literature on such plants is therefore widely dispersed.

However, when this literature is collated it is seen that specialized habitats such as metal-contaminated areas are extremely valuable to evolutionary and ecological studies. Their value stems essentially from their simplicity and this is a consequence of three main features.

Firstly, the vegetation of such areas is influenced by one overriding factor, namely metal concentration. Undoubtedly other factors are also of importance but such factors and interactions are far more easily defined when the major determinant is clearly apparent. Secondly, the habitats are usually spatially distinct and clear cut. This is particularly true of areas resulting from mining activities. Thirdly, since the major factor involved is an edaphic one it is relatively constant and varies only slowly with time.

It is perhaps not surprising, therefore, that such areas have already proved invaluable in studies of evolution. The sequence and pattern of genetic change responsible for colonization of metal-contaminated areas is a unique record of natural selection in action. It is clear that studies of metal tolerance may be of equal value in clarifying ecological problems particularly when they impinge on other areas such as evolution, taxonomy, plant distribution and physiological processes.

A pointer towards these interactions has been made in this paper, mainly by reference to work which has been carried out without the specific aim of achieving an integrated view.

Certain problems of ecology can often be better analysed in environments that are highly specialized or even man-made, than in natural, more complex habitats.

Acknowledgements

Many workers have provided us with advanced information of their results, either by personal communication or by drawing our attention to unpublished theses. Their names appear in the text, but we would like to thank them for their help. We are also grateful to all those who have helped us at one time or another to locate or compile many of the references.

REFERENCES

Allen, R. and Sheppard, P. M. (1971). *Proc. R. Soc. B* **177**, 177–196. Copper tolerance in some Californian populations of the Monkey Flower, *Mimulus guttatus*.

Anonymous (1959). *Horizon* **1**, 35–39. A flower that led to a copper discovery.

Antoine, A. (1965). *Expl Cell Res.* **40**, 570–584. La resistance de la levure aux ions cuivriques. III. *Saccharomyces cerevisiae*, Yeast foam, nature de deux formes resistantes.

Antonovics, J. (1966). "The genetics and evolution of differences between closely adjacent plant populations with special reference to heavy metal tolerance." Ph.D. Thesis, University of Wales.

Antonovics, J. (1968a). *Heredity, Lond.* **23**, 219–238. Evolution in closely adjacent plant populations. V. Evolution of self-fertility.

Antonovics, J. (1968b). *Heredity, Lond.* **23**, 507–524. Evolution in closely adjacent plant populations. VI. Manifold effects of gene flow.

Antonovics, J. and Bradshaw, A. D. (1970). *Heredity, Lond.* **25**, 349–362. Evolution in closely adjacent plant populations. VIII. Clinal patterns in *Anthoxanthum odoratum* across a mine boundary.

Antonovics, J., Lovett, J. and Bradshaw, A. D. (1967). *In* "Isotopes in Plant Nutrition and Physiology", pp. 549–567. Vienna I.A.E.A./F.A.O. Symposium. The evolution of adaptation to nutritional factors in populations of herbage plants.

Aomine, S., Kawasaki, H. and Inoue, K. (1967). *Soil Pl. Fd, Tokyo* **13**, 186–188. Retention of mercury by soils. I. Mercury residues of paddy and orchard soils.

Archer, J. (1964). "Zinc tolerance in *Agrostis tenuis*, *Agrostis stolonifera*, and their hybrids." B.Sc. Dissertation, University College of North Wales.

Ashida, J. (1965). *A. Rev. Phytopathol.* **3**, 153–174. Adaptation of fungi to metal toxicants.

Ashida, J. and Nakamura, H. (1959). *Pl. Cell Physiol., Tokyo* **1**, 71–79. Role of sulphur metabolism in copper-resistance in yeast.

Ashworth, L. J. (1965). Personal communication. Agric. Res. Service, U.S. Cotton Res. Sta., Shafter, Calif.

Ashworth, L. J. and Amin, J. V. (1964). *Phytopathology* **54**, 1459–1463. A mechanism for mercury tolerance in fungi.

Auquier, P. (1964). *Bull. Soc. r. Bot. Belg.* **97**, 99–130. Les Festuca des terrains calaminaires de la Wallonie Septentrionale.

Bailey, F. J. (1889). "Queensland Flora." Melbourne.

Bailey, F. M. (1898). *Bot. Zbl.* **76**, 104. The copper plant.

Bainbridge, A. (1969). *Agriculture, Lond.* **76**, 97–98. Mercury tolerance in leaf spot of oats.

Barker, J. A. (1967). Personal communication. Department of Agricultural Botany, University College of North Wales, Bangor, Caernarvonshire, U.K.

Bateman, W. G. and Wells, L. S. (1917). *J. Am. chem. Soc.* **39**, 811–819. Copper in the flora of a copper tailing region.

Baumann, A. (1885). *Landwn VersStnen* **31**, 1–53. Das Verhalten von Zinksalzen gegen Pflanzen und im Boden.

Baumeister, W. (1954). *Ber. dt. bot. Ges.* **67**, 205–213. Uber den Einfluss des Zinks bei *Silene inflata* Smith. I.

Baumeister, W. (1967). *Angew. Bot.* **40**, 185–204. Schwermetall-Pflanzengesellschaften und Zinkresistenz einiger Schwermetallpflanzen.

Baumeister, W. and Burghardt, H. (1956). *Ber. dt. bot. Ges.* **69**, 161–168. Uber den Einfluss des Zinks bei *Silene inflata* Smith. II. CO_2 Assimilation und Pigmentgehalt.

Beck, J. V. (1967). *Biotechnol. Bioeng.* **9**, 487–497. The role of bacteria in copper mining operations.

Beck, J. V. and Elsden, S. R. (1958). *J. gen. Microbiol.* **19**, i. Isolation and some characteristics of an iron-oxidising bacterium.

Bertrand, G. and Andreitcheva, M. (1963). *C.R. hebd. Seanc. Acad. Sci., Paris* **197**, 1374–1376. Sur la teneur comparée en zinc des feuilles vertes et des feuilles étiolées.

Bhatia, C. R. and Narayanan, K. R. (1965). *Genetics, Princeton* **52**, 577–581. Genetic effects of ethyl methanesulfonate in combination with copper and zinc ions on *Arabidopsis thaliana.*

Biebl, R. (1947a). *Sber. Akad. Wiss. Wien, Abt. I.* **155**, 145–157. Die resistenz gegen Zink, Bor und Mangan als Mittel zur Kennzeichnung verschiedener pflanzlicher Plasmasorten.

Biebl, R. (1947b). *Ost. bot. Z.* **94**, 61–73. Uber die gegensatzliche Wirkung der Spurenelemente Zink und Bor auf die Blattzellen von *Mnium rostratum.*

Biebl, R. (1950). *Protoplasma* **39**, 18–259. Uber die Resistenz pflanzlicher Plasmen gegen Vanadium.

Birrell, K. S. and Wright, A. C. S. (1945). *New Zealand Journ. Sci. and Technol.*, **27A**, 72–76. A serpentine soil in New Caledonia.

Booth, G. H. and Mercer, S. J. (1963). *Nature, Lond.* **199**, 622. Resistance to copper of some oxidising and reducing bacteria.

Bradshaw, A. D. (1952). *Nature, Lond.* **169**, 1098. Populations of *Agrostis tenuis* resistant to lead and zinc poisoning.

Bradshaw, A. D. (1959). *New Phytol.* **58**, 208–227. Population differentiation in *Agrostis tenuis* Sibth. I. Morphological differentiation.

Bradshaw, A. D., McNeilly, T. S. and Gregory, R. P. G. (1965). *In* "Ecology and the Industrial Society", *Brit. Ecol. Soc. Symp.* **5**, 327–343. Industrialization, evolution and the development of heavy metal tolerance in plants.

Bradshaw, A. D., Antonovics, J., Khan, M. S. and Walley, K. (1969). *Int. Bot. Cong. XI. Seattle, Abstracts*, p. 21. The importance of extreme selection pressures in evolution.

Bradshaw, A. D., Chadwick, M. J., Jowett, D., Lodge, R. W. and Snaydon, R. W. (1960). *J. Ecol.* **48**, 631–637. Experimental investigations into the mineral nutrition of several grass species. III. Phosphate level.

Braun-Blanquet, J. (1951). "Pflanzensoziologie", 2nd Edit. Springer, Wien.

Brock, J. D. (1969). *In* "Microbial growth". *Symp. Soc. Gen. Microbiol. 19th*, pp. 15–41. Microbial growth under extreme conditions.

Bröker, W. (1963). *Flora, Jena,* B. **153**, 122–156. Genetisch-physiologische Untersuchungen uber die Zinkvertraglichkeit von *Silene inflata Sm.*

Bryner, L. C. and Anderson, R. (1957). *Ind. Engng. Chem. ind. (int.) Edn.* **49**, 1721–1724. Microorganisms in leaching sulphide materials.

Bryner, L. C. and Jameson, A. K. (1958). *Appl. Microbiol.* **6**, 281–287. Microorganisms in leaching sulfide minerals.

Bryner, L. C., Beck, J. V., Davis, D. B. and Wilson, D. G. (1954). *Ind. Engng. chem. ind. (int.) Edn.* **46**, 2587–2592. Microorganisms in leaching sulfide minerals.

Buck, L. J. (1949). *Jl N.Y. bot. Gdn.* **50**, 265–269. Association of plants and minerals.

Cannon, H. L. (1957). *Bull. U.S. geol. Surv.* No. 1030-M, 399–516. Description of Indicator Plants and Methods of Botanical Prospecting for Uranium Deposits on the Colorado Plateau.

Cannon, H. L. (1960a). *Science, N.Y.* **132**, 591–598. Botanical prospecting for ore deposits.

Cannon, H. L. (1960b). *Bull. U.S. geol. Surv.* No. 1085A. The development of botanical methods of prospecting for uranium ore in the Colorado Plateau.

Chatterjee, B. D., Barua, A. D. and Banerjee, P. L. (1968). *Indian J. med. Res.* **56**, 395–401. Mercury resistance of *Staph. pyogenes* isolated from different sources of a hospital: its relationship to phage types and antibiogram.

Chester, V. E. (1965). *Proc. R. Soc. Ser. B.* **162**, 555–566. The role of calcium in the increase of flocculence of yeast growing in the presence of copper.

Chikishev, A. G. (1965). "Plant indicators of soils, rocks and sub-surface water." Consultants Bureau, New York.

Cholak, J., Schaffer, L. J. and Yeager, D. (1968). *Am. ind. Hyg. Ass. J.* **29**, 562–568. The air transport of lead compounds in automobile exhaust gases.

Clarke, O. M. Jnr. (1953). *Econ. Geol.* **48**, 39–45. Geochemical prospecting for copper at Ray, Arizona.

Coackley, A. and Dawson, M. (1966). "Cross tolerance in *Rumex acetosa*". B.Sc. Dissertation, University College of North Wales.

Cole, M. M. (1965). *Proc. Commonw. Min. Metall. Congr. 8th,* **6**, 1429–1458. The use of vegetation in mineral exploration in Australia.

Cole, M. M., Provan, D. M. J. and Tooms, J. S. (1968). *Trans. Instn Min. Metall.* **77**, 81-104. Geobotany, biogeochemistry and geochemistry in mineral exploration in the Bulman-Waimura Springs area, Northern Territory, Australia.

Craig, G. C. (1970). Personal communication. Department of Biology, University of Stirling, Scotland, U.K.

Da Costa, E. W. B. and Kerruish, R. M. (1964). *Forest. Prod. J.* **14**, 106–112. Tolerance of *Poria* species to copper based wood preservatives.

Davis, B. D. (1958). *In* "Drug Resistance in Micro-organisms", Wolstenholme, G. E. W. and O'Connor, C. M. (Eds). Churchill, London. Physiological (phenotypic) mechanisms responsible for drug resistance.

Dean, K. C., Dolezal, H. and Havens, R. (1968). A Progress Report, U.S. Department of the Interior, Bureau of Mines, Pittsburgh, Pa. Utilisation and Stabilisation of Solid Mineral Wastes.

Delas, J. (1963). *Agrochimica* **7**, 258–280. Toxicity of copper accumulated in the soil.

Demoulin, V., Lambinon, J., Maquinay, A. and Ramaut, J. L. (1967). *Bull. Jard. bot. nat. Belg.* **37**, 305–308. Teneur en zinc et en plomb de quelques Gasteronycetes des terrains calaminaires belges.

Doksopulo, E. P. (1961). Tbilisi, Izd-vo Tbilisskogo Univ. Nickel in rocks, soils, waters and plants adjacent to the talc deposits of the Chorchanskaya group.

Dorn, Von P. (1937). *Biologie* **6**, 11–13. Pflanzen als Anzeichnen fur Erzlagerstatten.

Drouineau, G. and Mazoyer, R. (1956). *Int. Congr. Soil Sci. 6th. Reports D*, pp. 419–421. Copper toxicity in soils.

Duvigneaud, P. (1958). *Bull. Soc. r. Bot. Belg.* **90**, 127–286. La vegetation du Katanga et de ses sols metalliferes.

Duvigneaud, P. (1959). *Bull. Soc. r. Bot. Belg.* **91**, 111–134. Plant "cobaltophytes" dans le Haut Katanga.

Duvigneaud, P. and Denaeyer-de Smet, S. (1960). *In* "Rapports de sol et de la vegetation", pp. 121–139. Masson, Paris. Influence des sols toxiques sur la vegetation. Action de certains metaux lourds du sol (cuivre, cobalt, manganese, uranium) sur la vegetation dans le Haut Katanga.

Duvigneaud, P. and Denaeyer-de Smet, S. (1963). *Bull. Soc. r. Bot. Belg.* **93**, 93–231. Cuivre et vegetation au Katanga.

Duvigneaud, P. and Timperman, J. (1959). *Bull. Soc. r. Bot. Belg.* **91**, 135–176. Etudes sur le genre *Crotalaria.*

D'Yakov, Y. T. (1963a). *Dokl. mosk. sel.-khoz. Akad. K.A. Timiryazeva* **83**, 267–272. Some physiological changes associated with the adaptation of *P. infestans* to fungicides.

D'Yakov, Y. T. (1963b). *Izv. timiryazev. sel. -khoz. Akad.* **2**, 39–46. Adaptation of the pathogen of potato late blight to copper acetate.

Dykeman, W. R. and de Sousa, A. S. (1966). *Can. J. Bot.* **44**, 871-878. Natural mechanisms of copper tolerance in a copper swamp forest.

Earley, E. B. (1943). *J. Am. Soc. Agron.* **35**, 1012–1023. Minor element studies with Soybeans. I. Varietal reactions to concentrations of zinc in excess of the nutritional requirement.

Ehrlich, H. L. (1962). *In* "Biogeochemistry of Sulphur Isotopes", M. L. Jenson (Ed.). National Science Foundation Symposium. Yale Univ., New Haven. pp. 153–168.

Ehrlich, H. L. (1963a). *Appl. Microbiol.* **11**, 15–19. Bacteriology of manganese nodules. I. Bacterial action on manganese on nodule enrichments.

Ehrlich, H. L. (1963b). *J. Bact.* **86**, 350–352. Micro-organisms in acid drainage from a copper mine.

Ehrlich, H. L. (1965). Personal communication. Dept. Biology, Rensselaer Polytechn. Inst., Troy, New York.

Ehrlich, H. L. and Fox, S. I. (1967). *Biotechnol. Bioeng.* **9**, 471–485. Environmental effects on bacterial copper extraction from low-grade copper sulfide ores.

Ernst, W. (1965a). *Abh. Landesmus. Naturk. Munster* **27**, (1), 54 pp. Okologische-Soziologische Untersuchungen der Schwermetall-Pflanzengesellschaften Mitteleuropas unter Einschluss der Alpen.

Ernst, W. (1965b). *Ber. dt. bot. Ges.* **78**, 205–212. Uber den Einfluss des Zinks auf die Keimung von Schwermetallpflanzen und auf die Entwicklung der Schwermetallpflanzengesellschaft.

Ernst, W. (1966). *Flora, Jena, B.* **156**, 301–318. Okologisch-soziologische Untersuchungen auf Schwermetallpflanzengesellschaften Südfrankreiches und des östlichen Harzvorlandes.

Ernst, W. (1967). *Excerpta bot., Sectio B.* **8**, 50–61. Bibliographie der Arbeiten uber Pflanzengesellschaften auf schwermetallhaltigen Boden mit Ausnahme des Serpentins.

Ernst, W. (1968a). *Mitt. flor.-soz. Arbgemein.* **13**, 263-268. Das Violetum calaminariae westfalicum, eine Schwermetallpflanzengesellschaft bein Blankenrode in Westfalen.

Ernst, W. (1968b). *Flora, Jena, B.* **158**, 95–106. Okologische Untersuchungen auf Pflanzengesellschaften unterschiedlich gestorter schwermetallreicher Boden in Grossbritannien.

Ernst, W. (1968c). *Ber. dt. bot. Ges.* **81**, 116–124. Zur Kenntnis der Soziologie und Okologie der Schwermetallvegetation Grossbritanniens.

Ernst, W. (1968d). *Physiologia Pl.* **21**, 323–333. Der Einfluss der Phosphatversorgung sowie die Wirkung von ionogenem und chelatisiertem Zink auf die Zink-und Phosphataufnahme einiger Schwermetallpflanzen.

Ernst, W. (1969a). *Ber. dt. bot. Ges.* **82**, 161–164. Zur Physiologie der Schwermetallpflanzensubselluläre Speicherungsorte des Zinks.

Ernst, W. (1969b). *Vegetatio* **18**, 393–400. Pollenanalytischer Nachweis eines Schwermetallrasens in Wales.

Foster, C. le N. (1894). "Ore and Stone Mining." Griffin, London.

Gadgil, R. L. (1969). *J. appl. Ecol.* **6**, 247–259. Tolerance of heavy metals and the reclamation of industrial waste.

Gambi, O. V. (1967). *G. Bot. Ital.* **101**, 59-60. First data on the histological localization of nickel in *Alyssum bertolonii* Desv.

Gibson, I. A. S. (1958). *E. Afr. agric. J.* **24**, 125–127. Phytotoxic effects of copper fungicides on acid soils.

Goldschmidt, U. M. (1954). "Geochemistry." Clarendon Press, Oxford.

Godwin, H. (1956). "The History of the British Flora." Cambridge University Press, London and New York.

Gorsline, G. W., Thomas, W. I. and Baker, D. E. (1964). *Crop Sci.* **4**, 207–210. Inheritance of P, K, Mg, Cu, B, Zn, Mn, Al and Fe concentrations by corn (*Zea Mays* L.) leaves and grain.

Gregory, R. P. G. (1965). "Heavy metal tolerance in Grasses." Ph.D. Thesis, University of Wales.

Gregory, R. P. G. and Bradshaw, A. D. (1965). *New Phytol.* **64**, 131–143. Heavy metal tolerance in populations of *Agrostis tenuis* Sibth. and other grasses.

Greenaway, W. and Cowan, J. W. (1970). *Trans. Br. mycol. Soc.* **54**, 127–138. The stability of mercury resistance in *Pyrenophora avenae.*

Gries, B. (1966). *Flora, Jena*, B. **156**, 271–290. Zellphysiologische Untersuchungen uber die Zinkresistenz bei Galmeiokotypen und Normalformen von *Silene cucubalus* Wib.

Griffith, J. W. (1919). *J. agric. Sci., Camb.* **9**, 366–395. Influence of mines upon land and livestock in Cardiganshire.

Harris, T. M. (1946). *New Phytol.* **45**, 50–55. Zinc poisoning of wild plants from wire netting.

Hassall, K. A. (1962). *Nature, Lond.* **193**, 90. A specific effect of copper on the respiration of *Chlorella vulgaris.*

Hawkes, H. E. (1948). *Circular No. 28. U.S. Department of Interior.* Annotated bibliography on papers on geochemical prospecting for ores.

Hawkes, H. E. (1957). *Bull. U.S. geol. Surv.* No. 1000-F, 225-352. Principles of geochemical prospecting.

Heimans, J. (1936). *Ned. kruidk. Archf.* **46**, 778-897. De Merkomst van de Zinkflora aan de Geul.

Heimans, J. (1960). *Publicaties van het natuurhistorisch Genootschap in Limburg.* **12**, (1960–61), 55–71. Taxonomic, phytogeographical and ecological problems round *Viola calaminaria* Lej.

Heimans, J. (1966). *Levende Nat.* **69**, 265–270. Het blauve Zinkviooltje van Westfallen.

Henwood, W. J. (1857). *Edin. New Phil. J.* **5**, 61–63. Notice of the copper turf of Merioneth.

Hilton, K. J. (1967). "The Lower Swansea Valley Project." Longmans, Green, London.

Hodgson, J. F., Lindsay, W. L. and Frierweiler, J. F. (1966). *Proc. Soil. Sci. Soc.*

Am. **30**, 723–726. Micronutrient cation complexing in soil solution. II. Complexing of zinc and copper in displaced solution from calcareous soils.

Hodgson, M. B. (1969). "Mercury resistance in ship-borne Enteromorpha." Hons. Thesis, Dept. Botany, Univ. Liverpool.

Hoekstra, W. G. (1964). *Fedn. Am. Socs. exp. Biol.* **23**, 1068–1076. Recent observations on mineral inter-relationships.

Hofsten, B. von (1962). *Expl. Cell. Res.* **26**, 606–667. The effect of copper on the growth of *Escherichia coli.*

Holmes, E. A. De B. (1965). "Root culture and measurement of lead tolerance in *Festuca ovina*." M.Sc. Thesis, University of Birmingham.

Holmes, R. S. (1964). Research report No. 23, Division of Soil and Fertiliser Investigations, Bureau of Plant Industry, Soils and Agricultural Engineering, U.S. Dept. Agriculture, Beltsville. The effect of liming on the availability to plants of zinc and copper.

Hooper, M. C. and Newton, L. (1935). *Minist. Agric. Fish. Rep.* **519**, (723). The effect of lead on plants with some suggestions for colonisation of Goginan mine.

Horn, P. and Wilkie, D. (1966). *Heredity, Lond.* **21**, 625-635. Selective advantage of the cytoplasmic respiratory mutant of *Saccharomyces cerevisiae* in a cobalt medium.

Horne, M. (1967). Personal communication. Department of Agricultural Botany, University College of North Wales, Bangor, Caerns. U.K.

Horsfall, J. G. (1956). "Principals of Fungicidal Action." Waltham, Massachusetts: Chronica Botanica.

Howard-Williams, C. (1969). "The ecology of *Becium homblei* Duvign. and Plancke." M. Phil. Thesis, University of London.

Howard-Williams, C. (1970). *J. Ecol.* **58**, 745–763. The ecology of *Becium homblei* in central Africa with special reference to metalliferous soils.

Humphreys, D. W. and Farnworth, J. (1964). "An investigation of the growth of *Minuartia verna* (a zinc tolerant species) on various soil types." B.Sc. Dissertation, University College of North Wales.

Jacobsen, W. B. G. (1967). *Kirkia* **6**, 63–80. The influence of the copper content of the soil on trees and shrubs of Molly South Hill, Mangula.

Jacobsen, W. B. G. (1968). *Kirkia* **6**, 259-277. The influence of the copper content of the soil on the vegetation at Silverside North, Mangula Area.

Jain, S. K. and Bradshaw, A. D. (1966). *Heredity, Lond.* **21**, 407–441. Evolutionary divergence among adjacent plant populations. I. The evidence and its theoretical analysis.

Jefferies, R. L., Laycock, D., Stewart, G. R. and Sims, A. P. (1969). *Brit. Ecol. Soc. Symp.* **9**, 281–308. *In* I. H. Rorison (Ed.) "Ecological aspects of the mineral nutrition of plants". The properties of mechanisms involved in the uptake and utilisation of calcium and potassium by plants in relation to an understanding of plant distribution.

Jenkins, A. E. and Winfield, R. J. (1964). "Zinc tolerance in *Holcus lanatus*." B.Sc. Dissertation, University College of North Wales.

Jensch, E. (1894). *Z. angew. Chem.* **7**, 14–15. Beitrage zur Galmeiflora von Oberschlesien.

Jones, O. T. (1922). *Mem. geol. Surv. spec. Rep. Miner. Resour. Gt. Br.* **XX**. 207 pp. Lead and zinc: the mining district of North Cardiganshire and West Montgomeryshire.

Jowett, D. (1958). *Nature, Lond.* **182**, 816-817. Populations of *Agrostis* spp. tolerant to heavy metals.

Jowett, D. (1959a). "Genecology of Heavy Metal Tolerance in Agrostis." Ph.D. Thesis, University of Wales.

Jowett, D. (1959b). *Nature, Lond.* **184**, 43. Adaptation of a lead tolerant population of *Agrostis tenuis* to low soil fertility.

Jowett, D. (1964). *Evolution, Lancaster, Pa.* **18**, 70–80. Population studies on lead tolerant *Agrostis tenuis*.

Kerin, Z. (1968). *Qualitas Pl. Mater. veg.* **15**, 372-379. Verunreinigungen von Gemuse aus Emissionen einer Bleihutte.

Khan, M. S. I. (1969). "The process of evolution of heavy metal tolerance in *Agrostis tenuis* and other grasses." M.Sc. Thesis, University of Wales.

Koch, K. (1932). *Wissenschaftliche Arbeiten des Bezirkskomitees fur Naturdenkmalpflege und Heimatschutz in Osnabruk* **1**, 91–115. Die Vegetationsverhaltnisse des Silberberges im Huggelgebiet bei Osnabruck.

Kruckeberg, A. R. (1954). *Ecology* **35**, 267–274. The ecology of serpentine soils III. Plant species in relation to serpentine soils.

Kruckeberg, A. R. (1964). *Am. Fern J.* **54**, 113–126. Ferns associated with ultramafic rocks in the pacific northwest.

Lackey, J. B. (1938). *Publ. Hlth Rep., Wash.* **53**, 1499–1507. The flora and fauna of surface waters polluted by acid mine drainage.

Lambinon, J. (1964). *Lejeunia*, **27**, 1–8. *Stereocaulon nanodes* Tuck. en Wallonie et en Rhenanie.

Lambinon, J. and Auquier, P. (1964). *Natura mosana* **16**, (4). La vegetation des terrains calaminaires de la Wallonie Septrentrionale et de la Rhenanie Aixoise. Types chorologiques et groupes ecologiques.

Lambinon, J., Maquinay, A. and Ramaut, J. L. (1964). *Bull. Jard. bot. Etat. Brux.* **34**, 273–282. La teneur en zinc de quelques lichens des terrains calaminaires Belges.

Lampe, W. and Klement, O. (1958). *Z. Mus. Hildesheim* **12**. Die Flechtenvegetation zwischen Oker und Leine in Raume von Hildesheim biz zum Harzrand.

Lange, O. L. and Ziegler, H. (1963). *Mitt. flor.-soz. ArbGemein.* **10**, 156–183. Der Schwermetallgehalt von Flechten aus dem *Acarosporetum sinopicae* auf Erzschalckenhalden des Harzes.

Lawrence, A. W. and McCarty, P. L. (1965). *J. Wat. Pollut. control Fed.* **37**, 392–406. The role of sulphide in preventing heavy metal toxicity in anaerobic treatment.

Lebrun, J., Noirfalaise, A., Heinemann, P. and Van Den Berghen, C. (1949). *Bull. Soc. r. Bot. Belg.* **82**, 105–207. Les Associations vegetales de Belgiques.

Lefebvre, C. (1967). *Bull. Soc. r. Bot. Belg.* **100**, 213–229. Etude de la position des populations d'*Armeria* calaminaires de Belgique et des environs d' Aix-la-Chappelle par rapport a des types alpines et maritimes d'*Armeria maritima* (Mill.) Willd.

Lefebvre, C. (1968). *Bull. Soc. r. Bot. Belg.* **102**, 5–11. Note sur un indice de tolerance au zinc chez des populations d'*Armeria maritima* (Mill.) Willd.

Lefebvre, C. (1969). Personal communication. Laboratoire de Génétique des Plantes Supérieures, Université Libre de Bruxelles, Belgium.

Lefebvre, C. (1970a). Personal communication. Laboratoire de Génétique des Plantes Supérieures, Université Libre de Bruxelles, Belgium.

Lefebvre, C. (1970b). *Evolution, Lancaster, Pa.* **24** (in press). Self-fertility in maritime and zinc mine populations of *Armeria maritima* (Mill.) Willd.

Le Riche, H. H. (1968). *J. agric. Sci., Camb.* **71**, 205–208. Metal contamination of soil in the Woburn market-garden experiment resulting from the application of sewage sludge.

Levin, D. A. and Kerster, H. W. (1967). *Evolution*, **21**, 679–687. Natural selection for reproductive isolation in *Phlox*.

Lewis, H. (1962). *Evolution, Lancaster, Pa.* **16**, 257–271. Catastrophic selection as a factor in speciation.

Linstow, O. Von (1929). *Abh. preuss. geol. Landesanst.* **114**. Bodenanzeigende Pflanzen.

Lohrmann, W. (1940). *Arch. Mikrobiol.* **11**, 329–367. Untersuchungen uber die antagonistische Wirkung von Magnesium gegenüber Bor und Quecksilber bei eingen Pilzen.

Lovering, T. S., Huff, L. C. and Almond, H. (1950). *Econ. Geol.* **45**, 493–514. Dispersion of copper from the San Manuel copper deposit, Pinal County, Arizona.

Lyon, G. L., Brooks, R. R., Peterson, P. J. and Butler, G. W. (1968). *Plant and Soil*, **29**, 225–240. Trace elements in a New Zealand serpentine flora.

McAllister, H. (1965). *Biol. J. Univ. St. Andrews* **10-11**. The ecology of an old rifle range target area with lead poisoned soil.

McBrian, D. C. H. and Hassall, K. A. (1965). *Physiologia Pl.* **18**, 1059–1065. Loss of cell potassium by *Chlorella vulgaris* after contact with toxic amounts of copper sulphate.

McHargue, J. S. and Roy, W. R. (1932). *Bot. Gaz.* **94**, 381–393. Mineral and nitrogen content of leaves of some forest trees at different times of the growing season.

McNeilly, T. S. (1965). "The evolution of copper tolerance in *Agrostis tenuis* (Sibth)." Ph.D. Thesis, University of Wales.

McNeilly, T. (1968). *Heredity, Lond.* **23**, 99–108. Evolution in closely adjacent plant populations. III. *Agrostis tenuis* on a small copper mine.

McNeilly, T. (1970). Personal communication. Department of Botany, University of Liverpool, Liverpool, U.K.

McNeilly, T. and Antonovics, J. (1968). *Heredity, Lond.* **23**, 205–218. Evolution in closely adjacent plant populations. IV. Barriers to gene flow.

McNeilly, T. and Bradshaw, A. D. (1968). *Evolution, Lancaster, Pa.* **22**, 108–118. Evolutionary processes in populations of copper tolerant *Agrostis tenuis* Sibth.

Malone, J. P. (1968). *Pl. Path.* **17**, 41–45. Mercury-resistant *Pyrenophora avenae* in Northern Ireland seed oats.

Malyuga, D. P. (1947). *Izvest. Akad. Nauk SSSR, Ser. Geograf. i. Geofiz.* **11**, 135–138. Chemical composition of soils and plants as indicators in prospecting for metals.

Malyuga, D. P. (1950). In "Report at the Conference on Trace Elements in the U.S.S.R." (*Ref. dokl. na. konf. po mikroelementam SSSR*) *Moscow*. On endemic plant diseases in districts of nickel ore deposits in the southern Urals.

Malyuga, D. P. (1964). "Biogeochemical methods of prospecting." (Authorised translation.) Consultants Bureau, New York.

Malyuga, D. P., Makashkina, N. S. and Makarova, A. I. (1959). *Geokhimiya* **5**. Biogeochemical investigations at Kadzharan, Armenian S.S.R.

Maquinay, A. and Ramaut, J. L. (1960). *Naturalistes Belg.* **41**, 265–273. La teneur en zinc des plantes du *Violetum calaminariae*.

Maquinay, A., Lamb, I. M., Lambinon, J. and Ramaut, J. L. (1961). *Physiologia Pl.* **14**, 284–289. Dosage du zinc chez un lichen calaminaire belge; *Stereocaulon nanodes* Tuck f. *tyroliense* (Nyl.) M. Lamb.

Marmo, V. (1953). *Econ. Geol.* **48**, 211–224. Biogeochemical investigations in Finland.

Minguzzi, C. and Vergano, O. e. (1948). *Atti. Soc. Tosc. Sci. Nat.* **55**. Il contennto di nichel nelle ceneri di *Alyssum bertoloni* Desv.

Murayama, T. (1961). *Mem. Ehime Univ.* IIB, **4**, 53–66. Studies on the metabolic pattern of yeast with reference to its copper resistance. IV. Characteristics in the tricarboxylic acid cycle.

Nestvetaylova, N. G. (1955). *Trudy Vses. aerogeol. Tresta* **1**, 146–152. Geobotanical investigations in the search for ore deposits.

Nicolls, O. W., Provan, D. M. J., Cole, M. M. and Tooms, J. S. (1965). *Trans. Instn. Min. Metall.* **74**, 695–799. Geobotany and geochemistry in mineral exploration in the Dugald River Area, Cloncurry District, Australia.

Noble, M., MacGarvie, Q. D., Hams, A. F. and Leafe, E. L. (1966). *Pl. Path.* **15**, 23–28. Resistance to mercury of *Pyrenophora avenae* in Scottish seed oats.

Noguchi, A. (1956). *Kumamoto J. Sci. Ser. B.* **2**, 239–257. On some mosses of *Merceya*, with special reference to the variation and ecology.

Noguchi, A. and Furuta, H. (1956). *J. Hattori bot. Lab.* **17**, 32–44. Germination of spores and regeneration of leaves of *Merceya ligulata* and *M. gedeana.*

Novick, R. P. and Roth, C. (1968). *J. Bact.* **95**, 1335–1342. Plasmid-linked resistance to inorganic salts in *Staphylococcus aureus.*

Old, K. M. (1968). *Trans. Br. mycol. Soc.* **51**, 525–534. Mercury tolerant *Pyrenophora avenae* in seed oats.

Passow, H., Rothstein, A. and Clarkson, T. W. (1961). *Pharmac. Rev.* **13**, 185–224. The general pharmacology of heavy metals.

Persson, H. (1948). *Revue bryol. lichen* **17**, 75–78. On the discovery of *Merceya ligulata* in the Azores with a discussion of the so-called "copper mosses".

Persson, H. (1956). *J. Hattori Bot. Lab.* **17**, 1–18. Studies in "copper mosses".

Peterson, H. B. and Monk, R. (1967). Vegetation and metal toxicity in relation to mine and mill wastes—an annotated bibliography. *Utah State Univ. Agr. Exp. Sta., Circ. No. 148.*

Peterson, P. J. (1969). *J. exp. Bot.* **20**, 863–875. The distribution of zinc-65 in *Agrostis tenuis* Sibth. and *A. stolonifera* L. tissues.

Pigott, C. D. and Walters, S. M. (1954). *J. Ecol.* **42**, 95–116. On the interpretation of the discontinuous distributions shown by certain British species of open habitats.

Poelt, J. (1955). *Verh. zool.-bot. Ges. Wien* **95**, 107–113. Flechten der Schwarzen Wand in der Grossarl.

Poore, M. E. D. (1955a). *J. Ecol.* **43**, 226–244. The use of phytosociological methods in ecological investigations. I. The Braun-Blanquet system.

Poore, M. E. D. (1955b). *J. Ecol.* **43**, 245–269. The use of phytosociological methods in ecological investigations. II. Practical issues involved in an attempt to apply the Braun-Blanquet system.

Poore, M. E. D. (1955c). *J. Ecol.* **43**, 606–651. The use of phytosociological methods in ecological investigations. III. Practical applications.

Poore, M. E. D. (1956). *J. Ecol.* **44**, 28–50. The use of phytosociological methods in ecological investigations. IV. General discussion of phytosociological problems.

Prat, S. (1934). *Ber. dt. bot. Ges.* **52**, 65–67. Die Erblichkeit der Resistenz gegen Kupfer.

Prat, S. and Komarek, K. (1934). *Zvlastni otisk ze sbornicku map.* VII, **8**, 1–16. Vegetace u medenych dolu.

Putwain, P. D. (1963). "Zinc tolerance in *Anthoxanthum odoratum*." B.Sc. dissertation, University College of North Wales.

Rabanus, A. (1931). *Angew. Bot.* **13**, 352–357. The toximetric testing of wood preservatives.

Ramaut, J. L. (1964). *Naturalistes belg.* **45**, 133–145. Un aspect de la pollution atmospherique: l'action des poussieres de zinc sur les soles et les vegetaux dans la region de Prayon.

Rankama, K. (1947). *Min. Metall., N.Y.* **28**, 282–284. Some recent trends in prospecting.

Rankama, K. (1954). In "Geochemical Methods for Exploration of Ore Deposits". Moscow, Il. The uses of trace elements in the solution of certain problems in applied geology.

Razzell, W. E. and Trussell, P. C. (1963). *Appl. Microbiol.* **11**, 105–110. Microbiological leaching of metallic sulfides.

Repp, G. (1963). *Protoplasma* **57**, 643–659. Die Kupferresistenz des Protoplasmas hoherer Pflanzen auf Kupfererzboden.

Richards, C. A. (1925). *Proc. Am. Wood-Preservers Assoc.* 1925, 18–22. The comparative resistance of eighteen species of wood destroying fungi to zinc chloride.

Riddell, J. E. (1952). *Prelim. Rep. Dep. Mines, Queb.* 269, 1–15. Anomalous copper and zinc values in trees in Holland Township, Gaspe Peninsula, North County.

Robinson, W. O., Lakin, H. W. and Reichen, L. E. (1947). *Econ. Geol.* **43**, 572–583. The zinc content of plants on the Friedensville zinc slime ponds in relation to biochemical prospecting.

Rosenfeld, I. and Beath, O. A. (1965). "Selenium-Geobotany, biochemistry, toxicity and nutrition." Academic Press, New York and London.

Ruhling, A. and Tyler, G. (1968). *Bot. Notiser* **121**, 321–342. An ecological approach to the lead problem.

Ruhling, A. and Tyler, G. (1969). *Bot. Notiser* **122**, 248–259. Ecology of heavy metals—a regional and historical study.

Rune, O. (1953). *Acta phytogeogr. suec.* **31**, 1–139. Plant life on serpentine and related rocks in the North of Sweden.

Russell, G. and Morris, O. P. (1970). *Nature, Lond.*, **228**, 288–289. Copper tolerance in the marine fouling alga, *Ectocarpus siliculosus*.

Russell, P. (1955). *Nature, Lond.* **176**, 1123–1124. Inactivation of phenyl mercuric acetate in groundwood pulp by a mercury resistant strain of *Penicillium roqueforti* Thom.

Schacklette, H. T. (1965). *Geol. Survey. Bull. 1198C. Washington.* Bryophytes associated with mineral deposits and solutions in Alaska.

Schatz, A. (1955). *Bryologist* **58**, 113–120. Speculations on the ecology and photosynthesis of "copper mosses".

Schmidt, W. E., Haag, H. P. and Epstein, E. (1965). *Physiol. Pl.* **18**, 860-869. Absorption of zinc by excised barley roots.

Schubert, R. (1952). "Die Pflanzengesellschaften der schwermetallhaltigen Boden des Ostlichen Harzvorlandes." Diss. Halle/S.

Schubert, R. (1953). *Wiss. Z. Martin-Luther-Univ. Halle-Wittenb.* **3**, 51–70. Die Schwermetallpflanzengesellschaften des ostlichen Harzorlandes.

Schubert, R. (1954a). *Wiss. Z. Martin-Luther-Univ. Halle-Wittenb.* **3**, 863-882. Zur Systematik und Pflanzengeographie der charakterpflanzen der Mitteldeutschen Schwermetallpflanzengesellschaften.

Schubert, R. (1954b). *Wiss. Z. Martin-Luther-Univ. Halle-Wittenb.* **4**, 99–120. Die Pflanzengesellschaften der Bottendorfer Hohen.

Schultz, A. (1912). *Jber. westf. ProvVer. Wiss. Kunst.* **40**, 209-227. Uber die auf schwermetallhaltigen Boden wachsenden Phanerogamen Deutschlands.

Schwanitz, F. and Hahn, H. (1954a). *Z. Bot.* **42**, 179–190. Genetischentwicklungsphysiologische Untersuchungen an Galmei pflanzen. I. Pflanzengrosse und Resistenz gegen Zinksulfat bei *Viola lutea* Hudson, *Alsine verna* L., und *Silene inflata* Sm.

Schwanitz, F. and Hahn H. (1954b). *Z. Bot.* **42**, 459–471. Genetischentwicklungsphysiologische Untersuchungen an Galmeipflanzen. II. Uber Galmeibiotypen bei *Linum catharticum* L., *Campanula rotundifolia* L., *Plantago lanceolata* L. und *Rumex acetosa* L.

Schwickerath, M. (1931). *Beitr. NatDenkmPflage.* **14**, 463–503. Das Violetum calaminariae der Zinkboden in der Umgebund Aachens. Eine Pflanzensoziologische Studie.

Seal, K. (1970). "The isolation and preliminary investigation of copper tolerance shown by strains of *Aspergillus serreus* and *Penicillium notatum*." M.Sc. Thesis, University of Wales.

Seno, T. (1962). *Japan. J. Genet.* **37**, 207–217. Genetical studies on copper resistance of *Saccharomyces cerevisiae.*

Sheridan, J. E., Tickle, J. H. and Chin, Y. S. (1968). *N.Z. J. Agric. Res.* **11**, 601–605. Resistance to mercury of *Pyrenophora avenae* (condial state *Helminthosporium avenae*) in New Zealand seed oats.

Shimazono, H. (1951). *J. Jap. For. Soc.* **33**, 393–397. The biochemistry of wood rotting fungi: the accumulation of oxalic acid.

Shimwell, D. (1967). (Manuscript.) Heavy metal plant communities in the Pennines.

Singewald, J. T. (1928). *Maryland Geol. Survey Repts.* **12**, 158–191. The chrome industry in Maryland.

Smith, J. M. (1966). *Am. Nat.* **100**, 637–650. Sympatric speciation.

Smith, R. A. H. and Bradshaw, A. D. (1970). *Nature, Lond.* **227**, 376–377. The reclamation of toxic metalliferous wastes.

Soane, B. D. and Saunder, D. H. (1959). *Soil Sci.* **88**, 322–330. Nickel and chromium toxicity of serpentine soils in Southern Rhodesia.

Somers, E. (1963). *Ann. appl. Biol.* **51**, 425–437. The uptake of copper by fungal cells.

Spanis, W. C., Munnecke, D. E. and Solberg, R. A. (1962). *Phytopathology* **52**, 455–462. Biological breakdown of two organic mercurial fungicides.

Spence, D. (1970). (Manuscript.) Scottish serpentine vegetation.

Spilling, C. and Thomas, D. (1964). "Zinc tolerance in *Rumex acetosa*." B.Sc. Dissertation, University College of North Wales.

Starkey, R. L. (1964). *In* "Principles and Applications in Aquatic Microbiology" (Henkelekian and Dondero, eds.). Discussion on paper by Ehrlich, H. L. (1964). John Wiley & Sons, Inc., New York.

Starkey, R. L. and Waksman, S. A. (1943). *J. Bact.* **45**, 509. Fungi tolerant to extreme acidity and high concentrations of copper sulfate.

Stebbins, G. L. (1942). *Madroño* **6**, 241–272. The genetic approach to problems of rare and endemic species.

Street, H. E. and Goodman, G. T. (1967). *In* "The Lower Swansea Valley Project", K. J. Hilton (Ed.), pp. 71–110. Longmans, Green, London. Revegetation techniques in the lower Swansea Valley.

Suchodoller, A. (1967). *Ber. schweitz. bot. Ges.* **77**, 266–308. Untersuchungen uber den Bleigehalt von Pflanzen in der Nahe von Strassen und uber die Aufnalime und Translokation von Blei durch Pflanzen.

Sudzuki-Hills, F. (1963). *Agr. Tec. Chile* **23/24**, 15–62. Relave de cobra y aquas de riego del Rio Cachapoal.

Swedish Royal Commission on Natural Resources (1967). *Oikos* Suppl. 9, 9–51. The Mercury Problem: Symposium concerning Mercury in the Environment, Stockholm, Sweden, 1966.

Taschenberg, E. F., Mack, G. L. and Gambrell, F. L. (1961). *J. agric. Fd. Chem.* **9**, 207–209. DDT and copper residues in a vineyard soil.

Taylor, J. (1953). *Phytopathology* **43**, 268–270. The effect of continual use of certain fungicides on *Phylaspora obtusa.*

Temple, K. L. and Le Roux, N. W. (1964). *Econ. Geol.* **59**, 647–655. Syngenesis of sulphide ores: desorption of adsorbed metal ions and their precipitation as sulphides.

Terui, G., Mochizuki, T., Irie, R. and Takano, M. (1960). *Technol. Rept. Osaka Univ.* **10**, 279–290. Mass adaptation of mold spores to acquire drug resistance in the course of germination.

Thyssen, S. von. (1942). *Beitr. angew. Geophys.* **10**, 35–84. Geochemical and botanical relationships in the light of applied geophysics.

Tonomura, K., Maeda, F. and Futai, F. (1968). *J. Ferment. Technol., Osaka* **46**, 685–692. Studies on the action of mercury-resistant micro-organisms on mercurials.

Tonomura, K., Nakagami, T., Futai, F. and Maeda, K. (1968). *J. Ferment. Technol., Osaka* **46**, 506–512. Studies on the action of mercury-resistant micro-organisms on mercurials.

Tonomura, K., Maeda, K., Futai, F., Nakagami, T. and Yamada, M. (1968). *Nature, Lond.* **217**, 644. Stimulative vaporization of phenylmercuric acetate by mercury-resistant bacteria.

Tooms, J. S. and Jay, J. R. (1964). *Econ. Geol.* **59**, 826–834. The role of the biochemical cycle in the development of copper/cobalt anomalies in the freely drained soils of the Northern Rhodesian Copperbelt.

Tsung-Shan, K. (1957). *Scientia sin.* **6**, 1105–1119. Skarn type of metamorphic copper ore deposits in Lower Yangtse.

Turner, R. G. (1967). "Experimental Studies on Heavy Metal Tolerance." Ph.D. Thesis, University of Wales.

Turner, R. G. (1969). *In* "Ecological Aspects of the Mineral Nutrition of Plants". (I. H. Rorison, ed.). *Brit. Ecol. Soc. Symp.* **9**, pp. 399–410. Heavy metal tolerance in plants.

Turner, R. G. (1970). *New Phytol.* **69**, 725–731. The subcellular distribution of zinc and copper within the roots of metal tolerant clones of *Agrostis tenuis* Sibth.

Turner, R. G. and Gregory, R. P. G. (1967). *In* "Isotopes in Plant Nutrition and Physiology", pp. 493–509. Vienna: I.A.E.A./F.A.O. Symposium. The use of radioisotopes to investigate heavy metal tolerances in plants.

Turner, R. G. and Marshall, C. (1971). *New Phytol.* **70**, 539–545. The accumulation of ^{65}Zn by root homogenates of zinc tolerant and non-tolerant clones of *Agrostis tenuis* Sibth.

Tuxen, R. (1937). *Mitt. flor.-soz. ArbGemein.* **3**. Die Pflanzengesellschaften Nordwestdeutschlands.

Url, W. (1955). *Sber. Akad. Wiss. Wien. Mathem.-naturw. Klasse Abt. I.* **164**, 207–230. Resistenz von Desmidiaceen gegen Schwermetallsake.

Url, W. (1956). *Protoplasma* **46**, 768–693. Über Schwermetall-, zumal Kupfer-resistenz einiger Moose.

Urquhart, C. (1971). *Heredity* **26**, 19–33. Genetics of lead tolerance in *Festuca ovina*.

Vernon, L. P., Magnum, J. H., Beck, J. V. and Shafia, F. M. (1960). *Archs. Biochem. Biophys.* **88**, 227–231. Studies on a ferrous-ion-oxidising bacterium. II. Cytochrome composition.

Vinogradov, A. P. (1955). *Tr. Biogeokhim. Lab. Akad. Nauk SSSR.* No. 3, 5–31. Elemental composition of organisms and D. I. Mendelleev's Periodic System.

Vogt, T. (1939). *K. norske Vidensk. Selsk. Skr.* **12**, 81–84. Chemical and botanical ore prospecting in the Røros area.

Vogt, T. (1942a). *K. norske Vidensk. Selsk. Skr.* **15**, 5–8. Geochemical and geobotanical ore prospecting. II. *Viscaria alpina* (L) G. Don.

Vogt, T. (1942b). *K. norske Vidensk. Selsk. Skr.* **15**, 21–24. Geochemical and geobotanical ore prospecting. III. Some notes on the vegetation at the ore deposits at Røros.

Vogt, T. and Braadlie, O. (1942). *K. norske Vidensk. Selsk. Skr.* **15**, 25-28. Geochemical and geobotanical ore prospecting. IV. Vegetation and soil at the ore deposits at Røros.

Vogt, T. and Bugge, J. (1943). *K. norske Vidensk. Selsk. Skr.* **16**, 51–54. Geochemical and geobotanical ore prospecting. VIII. Determination of copper in plants from the Røros area by quantitative X-ray analysis.

Vogt, T., Braadlie, O. and Bergh, H. (1943). *K. norske Vidensk. Selsk. Skr.* **16**, 55–58. Geochemical and geobotanical ore prospecting. IX. Determination of Cu, Zn, Pb, Mn and Fe in plants from the Røros area.

Von Rosen, G. (1964). *Hereditas* **51**, 89–134. Mutations induced by the action of metal ions in Pisum. II. Further investigations on the mutagenic action of metal ions and comparison with the activity of ionizing radiation.

Vose, P. B. and Randall, P. J. (1962). *Nature, Lond.* **196**, 85–86. Resistance to aluminium and manganese toxicities in plants related to variety and cation exchange capacity.

Wachsmann, C. (1961). *Dissertation en der Mathematisch-Naturwissenschaftlichen Fakultät der Westfalischen Wilhelms-Universität zu Munster in Referaten* **19**, 35–46. Wasserkulturversuche zur Wirkung von Blei, Kupfer und Zink auf die Gartenform und Schwermetallbiotypen von *Silene inflata* Sm.

Walker, R. B. (1954). *Ecology* **35**, 259–266. The ecology of serpentine soils. II. Factors affecting plant growth on serpentine soils.

Walley, K. A. (1967). "A study of gene flow in populations of *Agrostis tenuis*." B.Sc. Dissertation, University College of North Wales.

Warncke, E. (1968). *Bot. Tidsskr.* **63**, 358–368. *Marchantia alpestris* in Denmark.

Warren, H. V. and Delavault, R. E. (1948). *Geophysics*, **48**, 609–624. Biogeochemical investigations in British Columbia.

Warren, H. V. and Delavault, R. E. (1949). *Bull. geol. Soc. Am.* **60**, 531–560. Further studies in biogeochemistry.

Warren, H. V. and Delavault, R. E. (1950a). *Bull. geol. Soc. Am.* **61**, 123–128. Gold and silver content of some trees and horsetails in British Columbia.

Warren, H. V. and Delavault, R. E. (1950b). *Trans. Can. Inst. Min. Metall.* **53**, 236–270. A history of biochemical investigation in British Columbia.

Warren, H. V. and Delavault, R. E. (1960). *Trans. R. Soc. Can. Sect. IV* **54**, 11–20. Observations on the biogeochemistry of lead in Canada.

Warren, H. V. and Delavault, R. E. (1962). *J. Sci. Fd Agric.* **2**, 96–98. Lead in some food crops and trees.

Warren, H. V. and Howatson, C. H. (1947). *Bull. geol. Soc. Am.* **58**, 803–820. Biogeochemical prospecting for copper and zinc.

Warren, H. V., Delavault, R. E. and Fortescue, J. A. C. (1955). *Bull. geol. Soc. Am.* **66**, 229–238. Sampling in biogeochemistry.

Warren, H. V., Delavault, R. E. and Irish, R. I. (1949). *Proc. Trans. R. Soc. Can. 3rd Ser. Sect. IV* **43**, 119–138. Biogeochemical researches on copper in British Columbia.

Weston, R. L., Gadgil, P. D., Salter, B. R. and Goodman, G. T. (1965). *In* "Ecology and the Industrial Society". *Brit. Ecol. Soc. Symp.* **5**, 297–325. Problems of revegetation in the lower Swansea Valley, an area of extensive dereliction.

White, W. H. (1950). *Trans. Can. Inst. Min. Metall.* **53**, 243–246. Plant anomalies related to some British Columbia ore deposits.

Wild, H. (1968). *Kirkia* **7**, 1–71. Geobotanical anomalies in Rhodesia. I. The vegetation of copper-bearing soils.

Wild, H. (1970). *Kirkia* **7** supplement, 1–62. Geobotanical anomalies in Rhodesia. 3. The vegetation of nickel-bearing soils.

Wilkins, D. A. (1957). *Nature, Lond.* **180**, 37–38. A technique for the measurement of lead tolerance in plants.

Wilkins, D. A. (1960). *Rep. Scott. Pl. Breed. Stn.* 1960, 85–98. The measurement and genetic analysis of lead tolerance in *Festuca ovina.*

Williams, J. (1830). "Faunula Grustensis, being an outline of the natural contents of the Parish of Llanrwst." (See Mem. Geol. Survey Sp. Rept. Min. Res. Gt. Britain **23**, 59).

Williams, D. and Morgan, E. (1964). "Zinc tolerance in *Plantago lanceolata.*" B.Sc. Dissertation, University College of North Wales.

Williams, R. F. (1967). *J. Hyg. Camb.* **65**, 299–309. Mercury resistance and tetracycline resistance in *Staphylococcus aureus.*

Williams, S. T. (1970). Personal Communication. Department of Botany. University of Liverpool, Liverpool, U.K.

Wood, J. G. and Sibley, P. M. (1950). *Aust. J. scient. Res. Ser. B. Biol. Sci.* **3**, 14–27. The distribution of zinc in oat plants.

Woolhouse, H. W. (1970). *In* "Phytochemical Phylogeny", Phytochemical Soc. (Ed.) Academic Press, London. pp. 207-231. Environment and enzyme evolution in plants.

Yamamoto, T. (1963). *J. Inst. Polytech. Osaka Ag. Univ. Ser. D. Biol.* **14**, 159–165. Resistance of yeast (Saccharomyces cerevisiae) to high concentrations of strontium chloride.

Young, G. Y. (1961). *U.S. Dep. Agr. Forest Serv., Forest Prod. Lab. Rept. No. 2223*, 7 pp. Copper tolerance of some wood rotting fungi.

Ecological Implications of dividing Plants into Groups with Distinct Photosynthetic Production Capacities

CLANTON C. BLACK

Biochemistry Department, University of Georgia, Athens, Georgia, U.S.A.

I. Introduction

The productivity of photosynthetic organisms is a fundamental factor in ecological relationships since solar energy is the ultimate source of energy in biological systems. Within the last five years a new facet of plant biology directly related to the productivity of plants has been clearly recognized for the first time. Data on plant anatomy, plant physiology and plant biochemistry have converged with the recognition of distinct groups of plants with several distinct characteristics including photosynthetic production capacity. The data to support the concept of dividing plants into distinct groups have been collected by numerous research workers in laboratories throughout the world and span more than eighty years of effort.

One of the distinguishing characteristics of each plant group is the rate of net photosynthesis which in one major group of plants is two to threefold higher than the second major group of plants. In ecological relationships manifold consequences may arise from a two to threefold difference in primary productivity. It is the purpose of this article to present the data which support the concept of dividing plants into distinct groups and then to analyse some ecological implications and consequences of this division.

II. Anatomical Criteria

A. Leaf Anatomy

Over eighty years ago in the classical textbook, "Physiological Plant Anatomy", Professor G. Haberlandt lucidly described the basic leaf anatomy of plants with various photosynthetic capacities. His description was, ". . . each vascular bundle of the leaf is surrounded by a layer of . . . photosynthetic cells". This leaf arrangement was described in the following genera: *Cyperus*, *Saccharum*, *Pennisetum*, *Cynodon*, *Panicum*, *Andropogon*, *Danthonia* and *Spartina*, and referred to in dicotyledonous plants. He also noted in tissues of specific species within a genus (i.e. *Cyperus*) " . . . a vascular bundle enclosed in a sheath of large colourless cells". Thus within the same genus he noted that the cells surrounding the vascular bundle, the bundle sheath cells, could in different species either be photosynthetic or non-photosynthetic. Professor Haberlandt, however, was unaware of the other physiological and biochemical criteria such as a difference in net photosynthetic capacity between two species within the same genus which will be given in later sections of this paper.

These distinct anatomical features so clearly described and graphically presented by Professor Haberlandt and primarily representing the work of Schwendener, Volkens and Haberlandt were not widely known or utilized until the last four years. One notable exception to this statement was attempts by several workers to revise the systematics of Gramineae based on leaf anatomy (Prat, 1936; Stebbins, 1956; Brown, 1958). One of the basic types of leaf anatomy described by

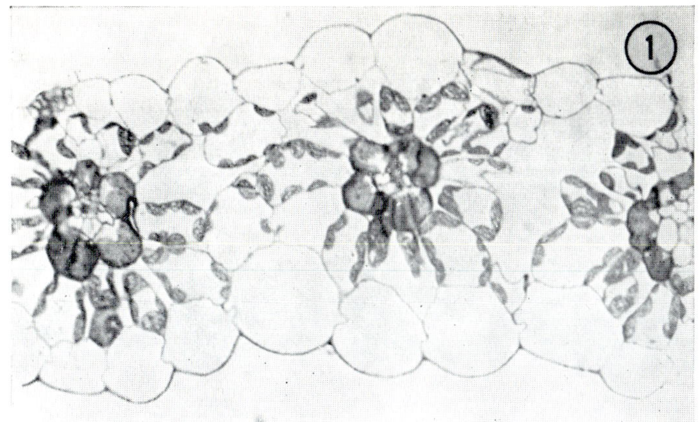

Fig. 1. Cross-section of a *Digitaria sanquinalis* (L.) Scop. leaf. Magnification—340 ×.

Haberlandt is illustrated in Fig. 1 with a leaf cross-section of *Digitaria sanquinalis* (L.) Scop. in which the leaf vascular tissues are surrounded by chloroplast-containing bundle sheath cells. In Fig. 2 the more common leaf anatomy in cross-sections is illustrated with *Triticum vulgare* L. These two figures illustrate the typical leaf anatomy in cross-section of a high photosynthetic capacity plant (Fig. 1), and a low photosynthetic capacity plant (Fig. 2).

B. Anatomy of Bundle Sheath Cells

A second unusual feature of the bundle sheath cells in high photosynthetic capacity plants was revealed in more detail by the electron microscope in that bundle sheath cells contain a remarkably high concentration of cellular organelles. Electron-micrograph pictures of bundle

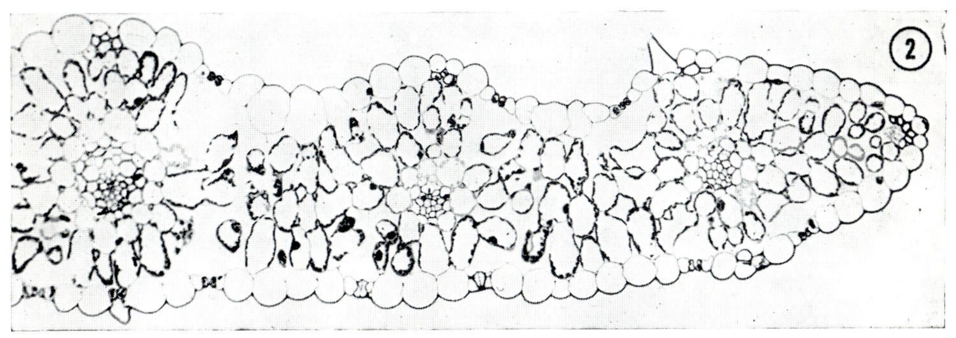

eaf. Magnification—340 ×.

ERRATUM

Page 89: the magnification factor in the legend to Fig. 2 should read 270x.

sheath cells are a graphic illustration of this characteristic. Figure 3 illustrates this remarkable concentration of organelles such as chloroplasts, mitochondria and peroxisomes in bundle sheath cells in a high photosynthetic capacity plant bermudagrass (*Cynodon dactylon* L.).

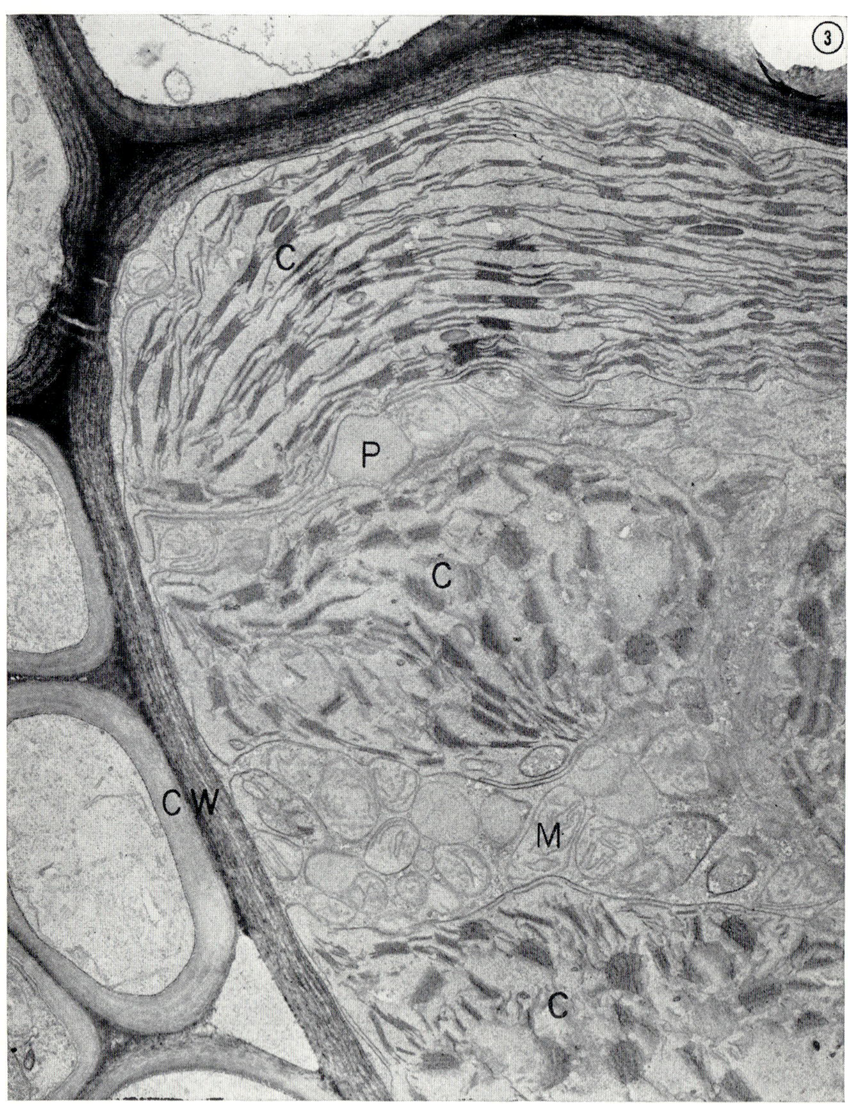

FIG. 3. Electron micrograph of a bundle sheath cell from a *Cynodon dactylon* L. leaf showing chloroplasts (C); mitochondria (M); peroxisome (P); cell wall (CW); and the general arrangement and concentration of organelles in a bundle sheath cell. Magnification—21 600.×

With fully developed leaves of high photosynthetic capacity plants, a high concentration of organelles in the bundle sheath cells seems to be a universal trait. In the work reported in Haberlandt's textbook (1914) it was recognized that the bundle sheath cells were dark green and other workers reported that the bundle sheath cell chloroplasts of maize (Kiesselbach, 1916) and sorghum (Rhoades and Carvalho, 1944) leaves were larger than mesophyll cell chloroplasts. However, the electron microscope revealed in preliminary work with maize something which went almost unnoticed for a decade (Hodge *et al.*, 1955), namely that the bundle sheath cells contained a high concentration of chloroplasts and "dense spherical bodies". In recent years the concentrating of organelles into bundle sheath cells has been clearly demonstrated in numerous genera including *Saccharum*, *Amaranthus*, *Atriplex*, *Sorghum*, *Aristida*, *Cynodon*, *Euchlaena*, *Panicum*, *Portulaca*, *Chloris*, *Digitaria*, *Eragrostis* and *Cyperus* (Johnson, 1964; Laetsch, 1968, 1969; Downton *et al.*, 1969; Black and Mollenhauer, 1971). Presently no exceptions have been reported to the statement that all high photosynthetic capacity plants, with the other characteristics described in this article, contain a high concentration of chloroplasts, mitochondria and other organelles in the bundle sheath cells. In fact, in our anatomical studies with bundle sheath cells often we cannot locate a vacuole (Fig. 3), or the vacuole is small and poorly defined simply because the bundle sheath cells are so densely packed with organelles (Black and Mollenhauer, 1971).

C. ULTRASTRUCTURE OF ORGANELLES

The chloroplasts and mitochondria of high photosynthetic capacity plants sometimes contain distinctive structural features (Rosado-Alberio *et al.*, 1968; Laetsch, 1968). Laetsch (1968) states that chloroplasts of *Amaranthus edulis* and *Atriplex lentiformis* have a highly developed peripheral reticulum composed of anastomosing tubules which are contiguous with the inner plastid membrane. This feature was observed in both bundle sheath and mesophyll cells of these high photosynthetic capacity plants. However, a well-developed peripheral reticulum in high photosynthetic capacity plants has not been demonstrated universally. Downton and Tregunna (1968) state that *Atriplex rosea* chloroplasts apparently do not contain this peripheral reticulum. Osmond *et al.* (1969) only found the peripheral reticulum in mesophyll cell chloroplasts of *Atriplex spongiosa* and furthermore state that detection of the reticulum is dependent upon the fixative employed. Thus more research is required to establish the distribution and validity of a highly developed chloroplast peripheral reticulum as a definite anatomical feature of all high photosynthetic capacity plants.

It is appropriate to note another point concerning the ultrastructure of the bundle sheath chloroplasts, namely, it was observed in early studies in maize and sugar-cane leaves that the bundle sheath cell chloroplasts lacked grana (Hodge *et al.*, 1955; Laetsch *et al.*, 1966); this phenomenon soon was correlated with other biochemical studies on these plants and the observation was made that bundle sheath cell chloroplasts lacked grana in plants which photo-reduced CO_2 via the C_4-dicarboxylic acid cycle. Subsequent research with a wide variety of high photosynthetic capacity plants has proved this idea to be invalid (Downton *et al.*, 1969; Laetsch, 1969; Osmond *et al.*, 1969; Black and Mollenhauer, 1971; Fig. 3).

The mitochondria of bundle sheath cells in high photosynthetic capacity plants generally are larger than mitochondria in mesophyll cells and the internal structure of the mitochondria are strikingly similar to the peripheral reticulum of the chloroplasts in Laetsch's studies (1969). Further research is needed on the ultrastructure of the mitochondria, but all studies support the general observation that mitochondria are concentrated in the bundle sheath cells (Fig. 3; Black and Mollenhauer, 1971). We have obtained biochemical data demonstrating that mitochondrial enzymes are ten times more active on a chlorophyll basis in bundle sheath cells than in mesophyll cells of *Digitaria sanquinalis*, which correlates well with the concentrating of mitochondria into bundle sheath cells observed in electron-microscopy studies (Edwards and Black, unpublished, 1970).

The distribution and ultrastructure of the peroxisome is unknown at this time and the biochemistry and physiology also is uncertain, as will be discussed in more detail in a later section. This very interesting organelle doubtless will be investigated in greater detail shortly and more specific conclusions may be reached.

For further reading, Laetsch (1969) recently presented a summary correlating chloroplast structure and photosynthetic carbon dioxide fixation pathways. The author agrees with his remark that, "The possession of . . . similar structural and functional features by unrelated taxa is a remarkable example of convergent evolution and many challenging questions can be asked about the selective forces of the environment responsible for these adaptations".

III. Physiological Criteria

When considering the following responses one should keep in mind that each environmental variable such as temperature, light intensity, oxygen and carbon dioxide pressure, and water, is involved in the physiology and metabolism of plants in a variety of fashions. For example carbon dioxide is removed from the atmosphere as a substrate of photosynthesis but concurrently it is released to the atmosphere as a product of photorespiration and respiration. Hence, net photosynthesis as measured by atmospheric carbon dioxide concentration is a direct function of these opposing processes which are in turn functions of numerous internal leaf reactions as well as external factors such as light intensity. Therefore, considerable effort is made to present the results of studies in which these variables were carefully considered.

A. TEMPERATURE OPTIMUM FOR PHOTOSYNTHESIS

One criterion distinguishing two groups of plants is the photosynthetic response to temperature which is illustrated in Fig. 4. The temperature optimum for net CO_2 uptake with high photosynthetic capacity plants is in the range of 30° to 45°C with CO_2 uptake decreasing rapidly at lower temperatures. In contrast, the photosynthesis temperature optimum for low photosynthetic capacity plants is in the broad range of 10° to 25°C with a decrease often noted as the temperature increases above 25°C. Miller (1960) initially reported a study comparing bentgrass (*Agrostis palustris*) with bermudagrass which illustrated these general responses. The most thorough comparative investigation has been in the laboratory of Professor Murata (1963, 1965; Fig. 4 was adapted by

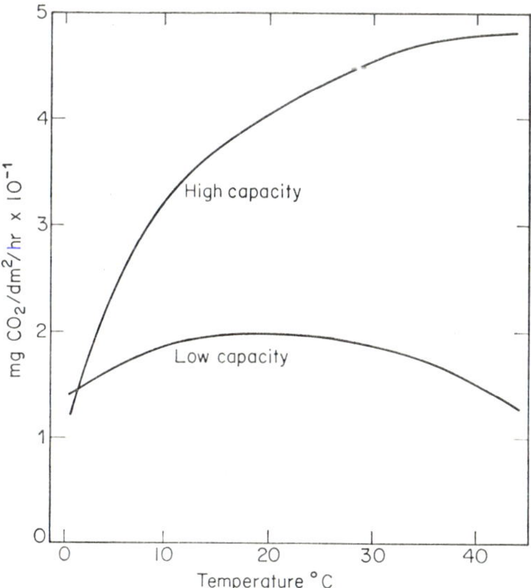

FIG. 4. General response curves to temperature for high and low photosynthetic capacity plants.

the author for these data) and several other laboratories have considered these general responses with other genera (El-Sharkawy and Hesketh, 1964; Cooper and Tainton, 1968; Chen *et al.*, 1970).

B. RESPONSE OF PHOTOSYNTHESIS TO LIGHT INTENSITY

One of the most striking characteristics of high photosynthetic capacity plants is the increasing CO_2 uptake concurrent with increasing light intensity. Figure 5 illustrates the typical response of CO_2 uptake

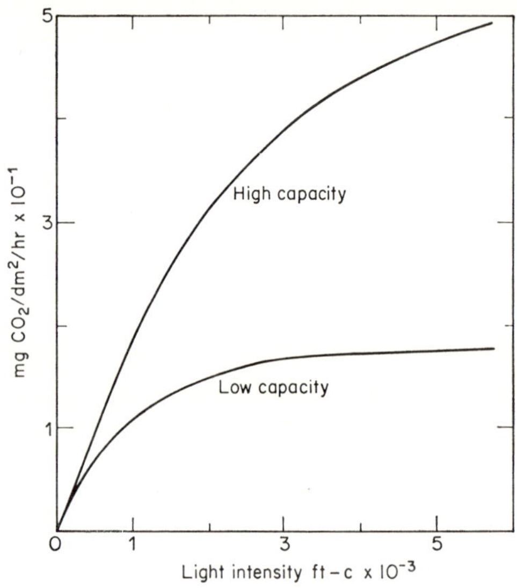

FIG. 5. General response curves to light intensity for high and low photosynthetic capacity plants.

with both a high and a low photosynthetic capacity plant as light intensity is increased toward full sunlight which often is reached at much higher intensities (10 000–12 000 foot candles).

The author suggests that readers utilize Fig. 5 as a reference point and clearly maintain it in mind. Figure 5 is the key-stone for the major thesis of this paper.

About a decade ago it was reported that intact leaves of maize, sugar-cane and sorghum had rates of net photosynthesis near 40 to 60 mg of CO_2/dm^2 leaf area/h (Moss *et al.*, 1961; Burr and Hartt, 1962; Hesketh and Musgrave, 1962; Hesketh and Moss, 1963; Hartt, 1965) which was several fold higher than CO_2 uptake by many other plants which is near 10–20 mg of $CO_2/dm^2/h$ (Rabinowitch, 1951; Böhning and Burnside, 1956). Subsequently maximum rates of net photosynthesis in the range of 50–80 mg of $CO_2/dm^2/h$ have been demonstrated in a variety of plants (Murata *et al.*, 1963, 1965; El-Sharkawy *et al.*, 1965, 1968; Brown *et al.*, 1966; Chen *et al.*, 1970; Fig. 5 was adapted from the data of Brown *et al.*, 1966, employing bermudagrass and orchardgrass (*Dactylis glomerata* L.) as the high and low capacity plant respectively).

Low photosynthetic capacity plants assimilate CO_2 within the range of 10–35 mg of $CO_2/dm^2/h$ and are saturated at light intensities in the range of 1–4000 foot candles (Rabinowitch, 1951).

C. OXYGEN PRESSURE RESPONSES

In 1920 Warburg reported that photosynthesis by *Chlorella* was inhibited at oxygen levels above atmospheric concentration (21%). An inhibition of photosynthesis by oxygen seems to be a paradox in that plants also produce oxygen during photosynthesis. McAlister and Myers (1940) confirmed these data and made the first observations on oxygen-inhibiting photosynthesis with higher plants in demonstrating that the atmospheric oxygen level was markedly inhibitory when compared with photosynthesis in 0·5% oxygen. They utilized wheat leaves and these observations were extended recently to other plants (Hesketh, 1963; Björkman, 1966; Björkman *et al.*, 1968; Forrester *et al.*, 1966; Hesketh, 1967; Downes and Hesketh, 1968). At atmospheric oxygen concentration (21%) photosynthesis is inhibited about 30–40%. Not only is photosynthesis inhibited, but the net result is to produce smaller plants as shown in the elegant studies of Björkman *et al.* (1968). They found that bean (*Phaseolus vulgaris* L.) seedlings and *Mimulus* grown in low oxygen (2·5 and 5% respectively) were 2·1 and 1·9 times larger than similar plants grown in 21% oxygen. However, all of these plants are low photosynthetic capacity plants.

Convincing data have been presented that high photosynthetic capacity plants *are not* inhibited by oxygen concentrations between about 1% and up to 100% (Forrester *et al.*, 1966; Björkman *et al.*, 1968; Downes and Hesketh, 1968). Again Björkman *et al.* (1968) convincingly demonstrated that growth of a high photosynthetic capacity plant, maize, was not affected by varying the oxygen concentration relative to atmospheric. Björkman's experimental results also were qualitatively similar at several CO_2 concentrations including atmospheric with either high or low photosynthetic capacity plants.

D. PHOTOSYNTHETIC CARBON DIOXIDE COMPENSATION CONCENTRATION

When photosynthetic organisms are placed in a closed, illuminated compartment, the CO_2 concentration in the compartment progressively decreases until no further change in concentration occurs with time, i.e. photosynthetic CO_2 uptake is equal to respiratory CO_2 evolution. This CO_2 concentration at which no further change occurs is defined as the photosynthetic CO_2 compensation concentration (Jackson and Volk, 1970).

High photosynthetic capacity plants have a photosynthetic CO_2 compensation concentration within the range of 0–10 ppm of CO_2 (Meidner, 1962; Moss, 1962; Forrester *et al.*, 1966; Tregunna, 1966; Tregunna and Downton, 1967; Downton and Tregunna, 1968; Krenzer and Moss, 1969; Moss *et al.*, 1969; Chen *et al.*, 1970). Meidner (1962) clearly demonstrated that maize could lower the CO_2 concentration in a closed compartment to a level lower than the detection limit of his equipment. The following genera, all of which are high photosynthetic capacity plants, contain species with CO_2 compensation concentrations of 10 ppm or less: *Zea*, *Saccharum*, *Cynodon*, *Paspalum*, *Echinochloa*, *Digitaria*, *Setaria*, *Chloris*, *Sorghum*, *Panicum*, *Andropogon*, *Eragrostis*, *Eleusine*, *Cyperus*, *Amaranthus*, *Gomphrena*, *Atriplex*, *Portulaca*, *Kochia* and *Salsola*.

Low photosynthetic capacity plants have photosynthetic CO_2 compensation concentrations in the range of 30–70 ppm of CO_2 (Miller and Burr, 1935; Gabrielsen, 1948; Decker, 1957, 1959; Moss, 1962; Forrester *et al.*, 1966; Tregunna and Downton, 1967; Zelitch, 1967; Downton and Tregunna, 1968; Krenzer and Moss, 1969; Moss *et al.*, 1969; Chen *et al.*, 1970.)

IV. Biochemical Criteria

A. Photosynthetic Carbon Dioxide Fixation Pathways

Prior to 1965–66 the reductive pentose phosphate cycle generally was accepted as the major process whereby photosynthetic organisms reduced CO_2 (Bassham and Calvin, 1957; Racker, 1957). In 1965 Kortschak and co-workers in Hawaii reported that malic, aspartic and phosphoglyceric acids were the earliest detectable labelled products of photosynthetic $^{14}CO_2$ fixation in sugar-cane leaves (Kortschak *et al.*, 1965). This unexpected observation was in sharp contrast to previous studies with other organisms in that the first labelled product of $^{14}CO_2$ fixation was almost invariably phosphoglyceric acid (Bassham and Calvin, 1957). In the sugar-cane leaf the earliest (0·6 sec) detectable labelled products, malic and aspartic acids, accounted for 80% of the fixed $^{14}CO_2$ and phosphoglyceric acid and was labelled much more slowly, 20% of the radioactivity in 15 sec. In contrast, 80% of the total radioactivity was found in phosphoglyceric acid in a 15-sec exposure of soybean leaves to $^{14}CO_2$ (Kortschak *et al.*, 1965). Clearly these studies on the sugar-cane leaf indicated that a major, previously unobserved, photosynthesis process occurred in sugar-cane whereas in other plants such as soybeans the reductive pentose phosphate cycle predominated.

Subsequently a group of Australian workers, Hatch, Slack and co-workers, confirmed the observations with sugar-cane leaves, extended these types of experimental observations to include specific grasses, sedges and dicotyledonous plants, and outlined a new cycle for photosynthetic CO_2 fixation, the C_4-dicarboxylic acid cycle (Hatch and Slack, 1966, 1967, 1968; Hatch *et al.*, 1967; Slack and Hatch, 1967; Osmond, 1967; Johnson and Hatch, 1968).

In outline this new cycle of photosynthetic CO_2 fixation involves the following reactions as proposed by Hatch and Slack.

I. Synthesis of phosphoenolpyruvate:

$$\text{Pyruvate} + \text{ATP} + \text{Pi} \xrightarrow[\text{Dikinase}]{\text{Pyruvate Pi}} \text{Phosphoenolpyruvate} + \text{AMP} + \text{Pyrophosphate}$$

II. Carboxylation of phosphoenolpyruvate:

$$\text{Phosphoenolpyruvate} + CO_2 \xrightarrow[\text{Carboxylase}]{\text{Phosphoenolpyruvate}} \text{Oxaloacetate} + \text{Pi}$$

III. Equilibration of oxaloacetate with malate and aspartate:

$$\text{Malate} \xleftrightarrow[\text{Dehydrogenase}]{\text{Malate}} \text{Oxaloacetate} \xleftrightarrow[\text{Aminotransferase}]{\text{Aspartate}} \text{Aspartate}$$

IV. Regeneration of pyruvate:

$$C_4\ \text{Acid} + \text{“2-carbon acceptor”} \xrightarrow{\text{“Transcarboxylase”}} \text{Pyruvate} + \text{Phosphoglycerate}$$

In the original experiments of Kortschak *et al.* (1965) oxaloacetate presumably was destroyed in the initial extraction procedure and thus was not detected. Hatch and Slack have partially substantiated the proposed C_4 cycle with subsequent data on enzyme activities and ^{14}C labelling patterns. The enzymes catalysing reactions I, II and III have been demonstrated in leaf extracts with activities high enough to support *in vivo* rates of CO_2 fixation. Short-time ^{14}C labelling studies indicate that carbon atom 4 of a C_4 acid is labelled initially and the next labelled compound is phosphoglyceric acid which is C-1 labelled (see reaction IV). Label then spreads to sugars. Thus, reaction IV *is postulated*—transcarboxylase activity has not been demonstrated to date nor has a "carbon acceptor" been identified.

In addition to the enzymes which are involved in the C_4 cycle of photosynthesis, the enzymes of the pentose cycle *also* are present in leaves of these plants (Hatch *et al.*, 1969; Osmond *et al.*, 1969; Slack, 1969; Slack *et al.*, 1969; Edwards *et al.*, 1970). Substantial levels of these enzymes are present in leaf extracts. Furthermore, the enzymes are located in specific cells within the *same* leaf (Edwards *et al.*, 1970). The bundle sheath cells (Figs 1 and 3) contain the enzymes to operate the pentose cycle and the mesophyll cells (Fig. 1) contain the enzymes (Reaction I, II and III) that partially operate the C_4 cycle (Edwards *et al.*, 1970).

In summary, a major new cycle of photosynthetic CO_2 fixation, the C_4-dicarboxylic acid cycle, has been proposed and partially substantiated in the last five years. Future research should establish the complete operation of the cycle, i.e. what is the "2 carbon acceptor" or is there a "transcarboxylation" as proposed in reaction IV. In addition, elucidation of the role(s) of two cycles of photosynthesis, in the same leaf, in adjacent cells, presents a very challenging research opportunity.

In 1965 Kortschak *et al.*, noted that sugar-cane is one of the most efficient plants for producing carbohydrates by photosynthesis. All plants which have given a positive test for the presence of the C_4 cycle are high photosynthetic capacity plants (Hatch *et al.*, 1967; Johnson and Hatch, 1968; Slack and Hatch, 1967; Black *et al.*, 1969; Chen *et al.*, 1969, 1970).

Low photosynthetic capacity plants appear to lack the key C_4 cycle enzymes such as pyruvate phosphate dikinase. Hence, the reductive pentose phosphate cycle appears to be the predominant cycle for CO_2 fixation in low photosynthetic capacity plants (Bassham and Calvin, 1957; Hatch *et al.*, 1967; Slack and Hatch, 1967; Black *et al.*, 1969; Chen *et al.*, 1970).

B. PHOTORESPIRATION

Photorespiration presents a paradox in that green plant photosynthesis closely involves the fixation of CO_2 and the evolution of O_2 but, when low photosynthetic capacity plants are illuminated, not only does photosynthesis occur but CO_2 evolution and O_2 uptake are stimulated by light. Decker in 1955–1959 firmly established that respiration of various plants increased following illumination (Decker, 1955, 1957, 1959a, b). Decker's pioneering research on photorespiration was largely ignored except in the laboratory of Professor G. Krotkov, where Decker's observations were confirmed and extended to demonstrate that some plant species, e.g. maize, did not exhibit photorespiration (Tregunna *et al.*, 1961, 1964; Forrester *et al.*, 1966 I, II). In recent years renewed interest in photorespiration was evident when it was noted that plants which did not exhibit photorespiration also contained the C_4 cycle of photosynthesis (El-Sharkawy *et al.*, 1967, 1968; Hatch *et al.*, 1967; Downton and Tregunna, 1968; Everson and Slack, 1968; Johnson and Hatch, 1968). Then evidence was presented by several workers that glycolate probably is the major substrate for photorespiration and that photorespiration could be controlled either with inhibitors or genetically (Zelitch, 1966, 1967, 1968; Moss, 1967, 1968; Zelitch and Day, 1968). Finally a new cellular organelle was isolated from plants, the peroxisome, and it was postulated that the peroxisome catalysed glycolate oxidation, hence the peroxisome appeared to be the site of photorespiration (Mollenhauer *et al.*, 1966; Tolbert *et al.*, 1968, 1969; Kisaki and Tolbert, 1969).

From these studies the hypothesis arose that a major difference between high and low photosynthetic capacity plants was the absence or presence of photorespiration. On the basis of this hypothesis the difference between the two curves in Fig. 5, as the illumination intensity increases, could be accounted for by the CO_2 evolved from glycolate oxidation. As it is now postulated, photorespiration clearly is a wasteful oxidation process which is not coupled to any useful form of energy conservation. Zelitch (1967) has calculated that with tobacco at 35°C as much as 60% of the CO_2 fixed is released through photorespiration. Since photorespiration increases with temperature, Zelitch's data can be extrapolated to postulate that the drop in net photosynthesis at 30°–40°C with low capacity plants (Fig. 4) is a manifestation of the increase in photorespiration as the temperature increases (Zelitch, 1967).

The author considers the hypothesis that photorespiration is a biochemical difference between the two groups of plants as tentative for the following reasons. The enzymes of glycolate metabolism, e.g. glycolate oxidase and catalase, are present in substantial quantities in leaf extracts from both high and low capacity plants. The substrate of photorespiration, glycolate, has been detected in leaves of both types of plants. Electron-microscopy studies (Fig. 3) clearly show a body, similar to a perixosome, in leaf cross-sections of both high and low photosynthetic capacity plants. Future research should resolve these inconsistencies.

Pending the outcome of such studies it is valid that at present all plants with easily detectable photorespiration are low photosynthetic capacity plants and plants without detectable photorespiration are high photosynthetic capacity plants. Thus the absence or presence of detectable photorespiration is a useful criterion.

V. Summation of Criteria for dividing Plants into Two Distinct Groups

A. General Conclusions

Table I summarizes the characteristics which distinguish two groups of higher plants. To the author's knowledge a plant which exhibits one

TABLE I

Characteristics which distinguish two groups of higher plants

Characteristic*	Photosynthetic capacity*†	
	High	Low
Highly developed bundle sheath cells in leaf cross-sections with an unusually high concentration of organelles	Present	Absent
Net rate of photosynthesis in full sunlight (10 000 – 12 000 foot candles)	40 – 80 mg of CO_2/ dm^2 leaf area/h	15 – 35 mg of CO_2/ dm^2 leaf area/h
Response of net photosynthesis to increasing light intensity	Difficult to reach saturation even at full sunlight	Saturation intensity reached in the range of 1000 – 4000 foot candles
Temperature optimum for photosynthesis	30 – 45°C	10 – 25°C
Photosynthetic CO_2 compensation concentration	0 – 10 ppm CO_2	30 – 70 ppm CO_2
Response of photosynthesis and growth to oxygen concentration	No detectable effect between 1 and 100% O_2	Inhibited as O_2 concentration increases above about 1% O_2
Major pathway(s) of photosynthetic CO_2 fixation	C_4-dicarboxylic acid and reductive pentose phosphate cycles	Reductive pentose phosphate cycle
Photorespiration	Not easily detectable	Present

* These characteristics apply to fully differentiated tissues only.
† All comparisons are at 21% oxygen, 0·03% carbon dioxide, or otherwise as near physiological conditions as possible.

TABLE II

Plant genera which possess both high and low photosynthetic capacity species

High photosynthetic capacity species	Low photosynthetic capacity species
Euphorbia maculata L.	*Euphorbia corollata* L.
Panicum milliaceum L.	*Panicum lindheimevi* Nash.
Cyperus rotundus L.	*Cyperus papyrus* L.
Atriplex rosea L.	*Atriplex hastata* L.

of these characteristics can, with confidence, be expected to exhibit the remaining characteristics. For example, there are no examples of plants which contain the enzymes of the C_4 cycle which are low photosynthetic capacity plants.

The reader should note clearly that we have considered *only* two major groups of plants. Probably other groups of plants exist with distinct characteristics such as intermediate CO_2 compensation concentration, or rates of photosynthesis which saturate at intensities below 1000 foot candles, or which fix CO_2 via other pathways, e.g. crassulacean metabolism plants. However, sufficient anatomical, physiological and biochemical data are not available on these plants to warrant proposing other distinct groups. It also is pertinent to note that this manuscript deals almost exclusively with angiosperms and we have particularly refrained from discussing gymnosperms and woody species. The primary

TABLE III

Monocotyledon plants with high or low photosynthetic capacity

High photosynthetic capacity monocotyledons	
Andropogon scoparium Michx.	*Leptochloa fusca* Kunth
Andropogon virginicus L.	*Panicum capillare* L.
Cenchrus ciliaris L.	*Panicum maximum* Jacq.
Cenchrus echinatus L.	*Panicum virgatum* L.
Cynodon dactylon (L.) Pers.	*Paspalum dilatatum* Poir.
Cyperus esculentus L.	*Paspalum notatum* L.
Cyperus rotundus L.	*Pennisetum purpureum* Schum.
Dactyloctenium aegyptium (L.) Rich.	*Saccharum officinarum* L.
Digitaria pentzil Stent.	*Setaria italica* (L.) Beauv.
Digitaria sanquinalis (L.) Scop.	*Setaria viridis* (L). Beauv.
Echinochloa colonum L.	*Sorghum bicolor* (L.) Moench
Echinochloa crusgalli (L.) Beauv.	*Sorghum halepense* (L.) Pers.
Eleusine indica (L.) Gaerth	*Sorghum vulgare* Pers.
Eragrostis chloromelas Steud.	*Spartina* spp.
Eragrostis pilosa (L.) Beauv.	*Trichachne californica* (Benth.) Chase.
Heteropogon contortus (L.) Beauv.	*Trichachne insularis* (L.) Nees.
Leptochloa dubia (H.B.K.) Nees.	*Tripsacum dactyloides* L.
	Zea mays L.
Low photosynthetic capacity monocotyledons	
Agropyron repens (L.) Beauv.	*Lolium multiflorum* Lam.
Agrostis alba L.	*Medica mutica* Walt.
Avena sativa L.	*Oryza sativa* L.
Cyperus alternifolium gracilis L.	*Panicum commutatum* Schult.
Dactylis glomerata L.	*Phalaris arundinacea* L.
Eichhornia crassipes (Mart.) Solms.	*Phalaris canariensis* L.
Festuca arundinacea Schreb.	*Poa pratensis* L.
Hordeum vulgare L.	*Triticum aestivum* L.

reason for these omissions is the lack of sufficient data although some information is available and much of it was compiled recently by Larcher (1969).

B. Lists of Plant Species in Each Group

Table II is a list of genera in which species have been detected in both groups. One can easily conclude after examining the species listed in Tables II, III and IV that the division of plants into two groups in this manuscript does not follow the usual taxonomic criteria for plant classification.

Table III is a list of monocotyledon plants with high and low photosynthetic capacity and Table IV is a similar list for dicotyledon plants. These lists were compiled from the references cited in this manuscript and are only partially inclusive since these areas of research are very active and other genera and species are being tested regularly.

Table IV

Dicotyledon plants with high or low photosynthetic capacity

High photosynthetic capacity dicotyledons	
Amaranthus albus L.	*Gomphrena celsoides* L.
Amaranthus edulis Speg.	*Gomphrena globosa* L.
Amaranthus palmeri S. Wats.	*Kochia childsii* L.
Amaranthus retroflexus L.	*Kochia scoparia* (L.) Roth
Atriplex rosea L.	*Portulaca grandiflora* Hook.
Atriplex semibaccata R. Br.	*Portulaca oleracea* L.
Atriplex spongiosa F.v.M.	*Salsola kali* L.
Low photosynthetic capacity dicotyledons	
Ambrosia artemisifolia L.	*Ipomea* spp.
Ambrosia trifida L.	*Jacquemontia tamnifolia* (L.) Griseb.
Arachis hypogea L.	*Lactuca sativa* L.
Atriplex hastata L.	*Medicago arabica* (L.) Huds.
Beta vulgaris L.	*Oenothera* spp.
Brassica nigra (L.) Koch.	*Phaseolus vulgaris* L.
Cassia tora L.	*Phytolacca americana* L.
Chenopodium album L.	*Plantago lanceolata* L.
Cnicus benedictus L.	*Pueraria lobata* (Willd.) Ohwi.
Commelina communis L.	*Richardia scabra* L.
Crotalaria spectabilis Roth	*Sesbania exaltata* (Raf.) Cory
Datura stramonium L.	*Sida spinosa* L.
Daucus carota L.	*Spinacea oleracea* L.
Glycine max Merrill	*Stizolobium deeringianum* Bort.
Gossypium arboreum L.	*Vicia sativa* L.
Gossypium herbaceum L.	*Vigna sinensis* (L.) Endl.
Gossypium hirsutum L.	*Xanthium pennsylvanicum* Wallr.
Helianthus annuus L.	

VI. Ecological Implications and Consequences

A. Introduction

Once the concept of two distinct groups of higher plants is acknowledged, then numerous ecological implications and consequences can be derived. This section presents selected ecological applications deduced from the concept and its supporting data. In developing a hypothesis as a biochemical basis for plant competition, several of these ecological applications have been presented previously (Black *et al.*, 1969).

The plant characteristics discussed thus far lend the following traits to high photosynthetic capacity plants: (*a*) increased growth and vigour as the light intensity increases since the rate of net photosynthesis also increases and approaches a maximum value some two to threefold higher than low photosynthetic capacity plants; (*b*) increased growth and vigour as the temperatures rise above 15° – 20°C for similar reasons as in (*a*); (*c*) no inhibition of growth at normal oxygen concentrations and perhaps less inhibition of growth by the oxygen produced during plant photosynthesis; (*d*) less loss of reduced carbon, e.g. less glycolate oxidation, hence a saving of energy and substrates for other metabolic processes; (*e*) net fixation of CO_2 is less limited by low CO_2 concentrations, since high photosynthetic capacity plants are capable of fixing all of the detectable CO_2 in a closed chamber (Section IIID); (*f*) the presence of two active photosynthetic carboxylases and their associated enzymes in the same leaf of high photosynthetic capacity plants could result in a higher affinity for and more rapid uptake of CO_2. Phosphoenolpyruvate carboxylase has a kM for CO_2 of 4×10^{-4} M (Everson and Slack, 1968) and ribulose-1,5-diphosphate has a kM for CO_2 of about $4{\cdot}5 \times 10^{-4}$ M (Cooper *et al.*, 1969), so the activity of both carboxylases may allow the efficient uptake of atmospheric CO_2 and an efficient carboxylation of any internal leaf CO_2 which may be released by such processes as respiration or photorespiration; (*g*) with the literature currently available, it is difficult to understand the advantages of highly developed bundle sheath cells. One can infer that the unusual anatomy of leaves from high photosynthetic capacity plants results in more rapid rates of translocation and perhaps higher concentrations of translocates (Hartt and Kortschak, 1967; Hofstra and Nelson, 1969a, b; Moss and Rasmussen, 1969) which may both prevent a feedback type inhibition of photosynthesis by a product such as starch and furnish nutrients to non-photosynthetic portions of the plants such that high growth rates are facilitated.

B. PLANT COMPETITION

Competition among plants may depend upon many characteristics such as their capacity to extract nutrients or moisture from the soil, differential responses to temperature, morphology, seed viability, or a variety of other factors. However, we have proposed that competitive ability depends on the net capacity of a plant to assimilate CO_2 and use the photosynthate to extend its foliage or increase its size (Black *et al.*, 1969). Plants which fix net amounts of CO_2 at higher rates and with greater capacities have an initial advantage when competing with plants with lower net photosynthetic capacities. Recently we presented a general hypothesis that high photosynthetic capacity plants are very competitive and that plant competition could be explained on a biochemical basis (Black *et al.*, 1969). From the data presented earlier and in this manuscript, it is clear that the CO_2 fixing abilities and respiratory activities of plants are mediated through biochemical reactions which are now just beginning to be understood. Hence, it seems that as net CO_2 fixation in plants is elucidated a firmer biochemical basis for plant competition also will be elucidated.

C. PLANTS IN STRESS ENVIRONMENTS

High photosynthetic capacity plants appear to contain traits which would enable them to adapt and compete effectively in ecosystems containing various types of environmental stress. In the following instances "stress environments" are briefly described and correlated with the fact that high photosynthetic capacity plants often are located in these environments. The author uses the term "stress environment" in a functional manner and utilizes common descriptive terms rather than using a strict classification of climates such as Thornthwaite's (1933). Within each example of a stress environment it is also a valid question to ask what is the limiting factor(s) of plant growth or photosynthesis in that environment?

1. Tropics

From the initial anatomical studies (Haberlandt, 1914), the physiological experiments of Murata *et al.* (1963, 1965) and the biochemical experiments of Kortschak *et al.* (1965) and Hatch and Slack (1966), tropical ecosystems appeared to be likely areas in which to find high photosynthetic capacity plants. Indeed this is true, since these environments are subject to the temperature values and light intensity values depicted in Figs 4 and 5 which should favour high photosynthetic capacity plants, and low photosynthetic capacity plants may even be partially inhibited at the higher temperature values (Fig. 5). Thus the idea arose that high photosynthetic capacity plants are "plants of tropical origin". Numerous high photosynthetic capacity species in Tables II, III and IV are plants of tropical origin. However, other plants also of tropical origin, such as rice, are low photosynthetic capacity plants. The author doubts the validity of restricting high photosynthetic capacity plants to "plants of tropical origin" and will point out some inconsistencies in this idea in subsequent sections. In tropical environments, what is the rate limiting factor(s)? Laetsch (1969) has proposed that a combination of both aridity and high temperature would increase CO_2 diffusion resistance and thereby result in CO_2 levels limiting photosynthesis, and high photosynthetic capacity plants have evolved an efficient CO_2 fixation system as a result of this selection pressure. Certainly some tropical environments are subject to wet and dry seasons but some are not.

2. *Arid regions and deserts*

The first thought of a limiting factor in arid and desert regions is moisture although other factors such as low night temperatures must be considered, since there is evidence that this may be a determining factor in plant growth in specific environments (Hilliard and West, 1970). Arid and desert regions are associated with short periods of adequate moisture supply which often is followed by periods of high temperature and high light intensity. Seemingly these factors would favour high photosynthetic capacity plants which could have evolved the combination of abilities to respond to increasing light and temperature and to be about twice as efficient as low photosynthetic capacity plants in utilizing water (see Section VID). Several high photosynthetic capacity genera in Tables II, III and IV are associated with arid or desert ecosystems including *Atriplex*, *Portulaca*, and *Andropogon*.

3. *Mountains*

Some of the most interesting stress environments in which high photosynthetic capacity plants have been reported are alpine regions (Rabinowitch, 1951). These plants often exhibit rates of net photosynthesis near 100 mg of $CO_2/dm^2/h$ in atmospheres of 0·02 – 0·04% CO_2. Unfortunately, extensive data on the anatomy, physiology and biochemistry of alpine plants are not available, thus it is not even certain which primary pathway of CO_2 fixation(s) is present in these plants. Again high light intensity is common in this environment plus a limited growing season. However, here the temperatures do not commonly reach the optimum levels in Fig. 4 and, since snow and ice usually are melting, water probably is not a limiting factor. These plants with an unusually high photosynthetic capacity, e.g. *Gentiana algicola*, *Crepis montana*, *Geum montanum* and *Alchemilla alpina* are not definitely classified in a distinct group in this manuscript since insufficient data are available to warrant a definite conclusion. Much of the available data on plants in these ecosystems has been compiled by Tranquillini (1964) and Billings and Mooney (1968). Perhaps plants in these areas are ecotypes as one could imply from the elegant work on this subject by Hiesey, Clausen and co-workers. (This work is given in detail in the annual *Carnegie Institute Year Book*.)

4. Estuaries

Estuaries are among the world's most productive ecosystems. For example, the salt marshes of the eastern coast of the United States are highly productive areas in which the primary producing plant is *Spartina* (Smalley, 1959). On the basis of its leaf anatomy and a low CO_2 compensation concentration (Haberlandt, 1914; Krenzer and Moss, 1969) *Spartina* would be classified as a high photosynthetic capacity plant although the primary pathway of photosynthetic CO_2 fixation is uncertain. The limiting factor in this environment is unknown although one could speculate on the influence of excess salt on internal CO_2 and H_2O levels and resistances. Other factors such as high light intensity and temperature cannot be limiting in this ecosystem.

5. Agriculture

In agriculture numerous ecosystems are established in which high photosynthetic production plants have definite roles. Obviously maize, sugar-cane and *Sorghum* are high photosynthetic capacity plants which are widely utilized as a food source, and numerous grasses, e.g. *Paspalum*, *Cynodon*, *Digitaria*, are utilized for animal feed. Numerous agricultural ecosystems can be explained on the basis of high photosynthetic capacity plants being highly competitive (Black *et al.*, 1969). For example, a short-time ecosystem is established for three to six months in most row crops and cultural practices are such that low photosynthetic capacity crop plants (most crop plants have a low photosynthetic capacity) can be grown, but even in this situation a few plants become pests. The plant pests (weeds) are almost invariably high photosynthetic capacity plants including genera such as *Cynodon*, *Digitaria*, *Sorghum*, *Setaria*, *Cyperus* and *Amaranthus* (Black *et al.*, 1969; Holm, 1969). Low photosynthetic capacity crop plants cannot compete in such an ecosystem unless favoured by cultural practices. Thus, most weedy plants in row crops are high photosynthetic capacity plants.

The lack of a major weedy plant problem in permanent pastures in the southern United States is another ecosystem which illustrates this concept. Most of the major pastures in this area are introduced plants—*Paspalum notatum*, *Cynodon dactylon* and *Digitatia decumbens*. Once these plants are established, weedy plants do not present a problem. In fact, it is a common occurrence in agriculture to find large fields of these plants without a major infesting weedy plant. In this ecosystem, these high photosynthetic capacity plants do not allow other plants to become pests.

D. WATER REQUIREMENTS

One of the most important factors limiting plant growth is water supply. Plants which make efficient use of water may be expected to grow more in periods of moisture stress than those which do not. An extensive determination of the efficiency of water use by plants was made by Shantz and Piemeisel (1927). A comparison of their findings with the division of plants into two groups in this manuscript reveals a striking correlation. Table V gives the water requirements for plants grown in large metal containers.

It is readily seen that high photosynthetic capacity plants require about half as much water as low photosynthetic capacity plants to produce one unit of dry matter. Roots of sugar beet were harvested, making its water requirement appear lower. Roots of other species were not harvested.

The reason(s) for the coincidence of low water requirements with high photosynthetic capacity is not known, but probably reflects the high CO_2 uptake rate and concomitant high growth rate which would decrease the amount of water required to produce one gram of dry matter, unless the transpiration rate increased in proportion to CO_2 uptake. El-Sharkawy and Hesketh (1965) and Osmond *et al.* (1969) indicate that higher photosynthetic rates were not accompanied by a rate of transpiration higher than occurred in species with low photosynthetic rates. Thus, it appears that high photosynthetic capacity plants have lower water requirements than low photosynthetic capacity plants.

E. COMPARATIVE PLANT BIOLOGY

Kluyver and van Niel developed the concept that there is a unified plan of life, or comparative biochemistry, and that the biochemistry of each organism reflects this unity. This penetrating tool has been employed extremely successfully in many areas of biology. However, in plant biology, plants often are compared with animals or bacteria but

TABLE V

Water requirements for plant species at Akron, Colorado (data of Shantz and Piemeisel, 1927)

High photosynthetic capacity species	Water requirement*	Low photosynthetic capacity species	Water requirement*
MONOCOTYLEDONS		MONOCOTYLEDONS	
Panicum miliaceum L.	267	*Hordeum vulgare* L.	518
Sorghum spp.	304	*Triticum durum* Desf.	542
Zea mays L.	349	*Triticum aestivum* L.	557
Setaria italica (L.) Beauv.	285	*Avena sativa* L.	583
Sorghum sudanense (Piper) Stapf.	305	*Secale cereale* L.	634
Bouteloua gracilis (H.B.K.) Lag.	338	*Agropyron desertorum* Fisch.	678
		Bromus inermis Leyss.	977
		Oryza sativa L.	682
DICOTYLEDONS		DICOTYLEDONS	
Amaranthus graecizans L.	260	*Beta vulgaris* L.	377
Amaranthus retroflexus L.	305	*Chenopodium album* L.	658
Salsola kali L.	314	*Polygonum aviculare* L.	678
Portulaca oleracea L.	281	*Gossypium hirsutum* L.	568
		Solanum tuberosum L.	575
		Xanthium pennsylvanicum Wallr.	415
		Solanum triflorum Nutt.	487
		Solanum rostratum Dunal.	536
		Helianthus annuus L.	623
		Artemisia frigida Willd.	654
		Verbena bractiosa Lag. and Rodr.	702
		Brassica oleracea capitata L.	518
		Brassica rapa L.	614
		Brassica napus L.	714
		Ambrosia artemisifolia L.	912
		Citrullus vulgaris Schrad.	577
		Cucumis sativus L.	686
		Vigna sinensis Endl.	569
		Trifolium incarnatum L.	636
		Phaseolus vulgaris L.	700
		Medicago sativa L.	844

* Grams of water required to produce one gram of dry matter.

seldom plants with plants, e.g. studies of monocots versus dicots have not revealed any striking biochemical differences. Within the last five years, however, comparative studies of high photosynthetic capacity plants versus low photosynthetic capacity plants have been very informative as documented herein. At this time in the course of plant research the author advocates that persons interested in plant biology, not just photosynthesis, should reconsider their interests in plants and realize that this powerful tool of comparative biology is being successfully utilized in comparative photosynthesis and could be utilized effectively in many other areas of plant biology.

VII. Concluding Remarks

Some ecologists will object to this paper on the ground that it advocates a "one factor ecology" approach. Namely, that photosynthetic capacity is a dominating factor in ecology. The author sees little to be gained from this argument and simply states that solar energy is the major energy source in our biosphere and a plant which can add energy two to three times faster into an ecosystem than another plant must be considered when explaining or studying that ecosystem.

Another intractable problem in this manuscript is the necessity of extrapolation from laboratory data on metabolic processes to natural ecosystems in the field.

Acknowledgements

The author is grateful to the following associates for lively discussions and the sharing of thoughts: Drs H. R. Brown, T. M. Chen, G. E. Edwards, L. J. Laber, S. S. Lee, B. C. Mayne and H. H. Mollenhauer. However, responsibility for the manuscript and its deficiencies rests with the author.

References

Bassham, J. and Calvin, M. (1957). "The path of carbon in photosynthesis, p. 104." Prentice-Hall, Englewood Cliffs, N. J.

Billings, W. D. and Mooney, H. A. (1968). *Biol. Rev.* **43**, 481–529. The ecology of arctic and alpine plants.

Björkman, O. (1966). *Physiologia Pl.* **19**, 618–633. The effect of oxygen concentration on photosynthesis in higher plants.

Björkman, O., Hiesey, W. M., Nobs, M., Nicholson, F. and Hart, R. W. (1968). pp. 228–245. Carnegie Inst. Year Book. Carnegie Inst. of Washington, Washington, D.C. Effect of oxygen concentration on dry matter production in higher plants.

Black, C. C., Chen, T. M. and Brown, R. H. (1969). *Weed Sci.* **17**, 338–344. Biochemical basis for plant competition.

Black, C. C. and Mollenhauer, H. H. (1971). *Pl. Physiol.* **47**, 15–23. Structure and distribution of chloroplasts and other organelles in leaves with various rates of photosynthesis.

Böhning, R. H. and Burnside, C. A. (1956). *Am. J. Bot.* **43**, 557–561. The effect of light intensity on rate of apparent photosynthesis in leaves of sun and shade plants.

Brown, R. H., Blaser, R. E. and Dunton, H. L. (1966). *Proc. int. Grassland Congr.* **10**, 108–113. Leaf area index and apparent photosynthesis under various microclimates for different pasture species.

Brown, W. V. (1958). *Bot. Gaz.* **119**, 170–178. Leaf anatomy in grass systematics.

Burr, G. and Hartt, C. (1962). *Rep. Hawaiian Sug. Plrs' Ass. Exp. Stn.* p. **9**.

Chen, T. M., Brown, R. H. and Black, C. C., Jr. (1969). *Pl. Physiol.* **44**, 649–654. Photosynthetic activity of chloroplasts isolated from bermudagrass (*Cynodon dactylon* L.), a species with a high photosynthetic capacity.

Chen, T. M., Brown, R. H. and Black, C. C. (1970). *Weed Sci.* **18**, 399–403. CO_2 compensation concentration, rate of photosynthesis, and carbonic anhydrase activity of plants.

Cooper, J. P. and Tainton, N. M. (1968). *Herb. Abstr.* **38**, 167–176. Light and temperature requirements for the growth of tropical and temperate grasses.

Cooper, T. G., Filmer, D., Wishnick, M. and Lane, M. D. (1969). *J. biol. Chem.* **244**, 1081–1083. The active species of "CO_2" utilized by ribulose diphosphate carboxylase.

Decker, J. P. (1955). *Pl. Physiol.* **30**, 82–84. A rapid, post-illumination deceleration of respiration in green leaves.

Decker, J. P. (1957). *J. sol. Energy Sci. Engng* **1**, 30–33. Further evidence of increased carbon dioxide production accompanying photosynthesis.

Decker, J. P. (1959a). *Pl. Physiol.* **34**, 100–102. Comparative responses of carbon dioxide outburst and uptake in tobacco.

Decker, J. P. (1959b). *Pl. Physiol.* **34**, 103–106. Some effects of temperature and carbon dioxide concentration on photosynthesis of *Mimulus*.

Downes, R. W. and Hesketh, J. D. (1968). *Planta* **78**, 79–84. Enhanced photosynthesis at low oxygen concentrations: differential response of temperate and tropical grasses.

Downton, J., Berry, J. and Tregunna, E. B. (1969). *Science* **163**, 78–79. Photosynthesis: temperate and tropical characteristics within a single grass genus.

Downton, W. J. S. and Tregunna, E. B. (1968). *Can. J. Bot.* **46**, 207–215. Carbon dioxide compensation—its relation to photosynthetic carboxylation reactions, systematics of the Gramineae, and leaf anatomy.

Edwards, G. E., Lee, S. S., Chen, T. M. and Black, C. C. (1970). *Biochem. biophys. Res. Commun.* **39**, 389–395. Carboxylation reactions and photosynthesis of carbon compounds in isolated mesophyll and bundle sheath cells of *Digitaria sanquinalis* (L.) Scop.

El-Sharkawy, M. A. and Hesketh, J. D. (1964). *Crop. Sci.* **4**, 514–518. Effects of temperature and water deficit on leaf photosynthesis rates of different species.

El-Sharkawy, M. A. and Hesketh, J. D. (1965). *Crop Sci.* **5**, 517–521. Photosynthesis among species in relation to characteristics of leaf anatomy and CO_2 diffusion resistances.

El-Sharkawy, M. A., Loomis, R. S. and Williams, W. A. (1967). *Physiologia Pl.* **20**, 171–186. Apparent reassimilation of respiratory carbon dioxide by different plant species.

El-Sharkawy, M. A., Loomis, R. S. and Williams, W. A. (1968). *J. appl. Ecol.* **5**, 243–251. Photosynthesis and respiratory exchanges of carbon dioxide by leaves of the grain amaranth.

Everson, R. G. and Slack, C. R. (1968). *Phytochem.* **7**, 581–584. Distribution of carbonic anhydrase in relation to the C_4 pathway of photosynthesis.

Forrester, M. L., Krotkov, G. and Nelson, C. D. (1966). *Pl. Physiol.* **41**, 422–427; 428–431. Effect of oxygen on photosynthesis, photorespiration and respiration in detached leaves. I. Soybean; II. Corn and other monocotyledons.

Gabrielsen, E. K. (1948). *Nature* **161**, 138–139. Threshold value of carbon dioxide concentration in photosynthesis of foliage leaves.

Haberlandt, G. (1914). "Physiological Plant Anatomy" (translation). Macmillan, London.

Hartt, C. E. (1965). *Pl. Physiol.* **40**, 718–724. Light and translocation of ^{14}C in detached blades of sugarcane.

Hartt, C. E. and Kortschak, H. P. (1967). *Pl. Physiol.* **42**, 89–94. Translocation of ^{14}C in the sugarcane plant during the day and night.

Hatch, M. D. and Slack, C. R. (1966). *Biochem. J.* **101**, 103–111. Photosynthesis by sugarcane leaves. A new carboxylation reaction and the pathway of sugar formation.

Hatch, M. D. and Slack, C. R. (1967). *Archs Biochem. Biophys.* **120**, 224–225. The participation of phosphoenolpyruvate synthetase in photosynthetic CO_2 fixation of tropical grasses.

Hatch, M. D. and Slack, C. R. (1968). *Biochem. J.* **106**, 141–146. A new enzyme for the interconversion of pyruvate and phosphopyruvate and its role in the C_4 dicarboxylic acid pathway of photosynthesis.

Hatch, M. D., Slack, C. R. and Bull, T. A. (1969). *Phytochem.* **8**, 697–706. Light-induced changes in the content of some enzymes of the C_4-dicarboxylic acid pathway of photosynthesis and its effect on other characteristics of photosynthesis.

Hatch, M. D., Slack, C. R. and Johnson, H. S. (1967). *Biochem. J.* **102**, 417–422. Further studies on a new pathway of photosynthetic carbon dioxide fixation in sugarcane and its occurrence in other plant species.

Hesketh, J. D. (1963). *Crop Sci.* **2**, 493–496. Limitations to photosynthesis responsible for differences among species.

Hesketh, J. D. (1967). *Planta* **76**, 371–374. Enhancement of photosynthetic CO_2 assimilation in the absence of oxygen, as dependent upon species and temperature.

Hesketh, J.D. and Moss, D. N. (1963). *Crop Sci.* **3**, 107–110. Variation in the response of photosynthesis to light.

Hesketh, J. D. and Musgrave, R. B. (1962). *Crop Sci.* **2**, 311–315. Photosynthesis under field conditions. IV. Light studies with individual corn leaves.

Hilliard, J. H. and West, S. H. (1970). *Science* **168**, 494–496. Starch accumulation associated with growth reduction at low temperatures in a tropical plant.

Hodge, A. J., McLean, J. D. and Mercer, F. V. (1955). *J. biophys. biochem. Cytol.* **1**, 605–614. Ultrastructure of the lamellae and grana in the chloroplast of *Zea mays*.

Hofstra, G. and Nelson, C. D. (1969a). *Can. J. Bot.* **47**, 1435–1442. The translocation of photosynthetically assimilated ^{14}C in corn.

Hofstra, G. and Nelson, C. D. (1969b). *Planta* **88**, 103–112. A comparative study of translocation of assimilated ^{14}C from leaves of different species.

Holm, L. (1969). *Weed Sci.* **17**, 113–118. Weed problems in developing countries.

Jackson, W. A. and Volk, R. J. (1970). *A. Rev. Pl. Physiol.* **21**, 385–432. Photorespiration.

Johnson, H. S. and Hatch, M. D. (1968). *Phytochem.* **7**, 375–380. Distribution of the C_4 dicarboxylic acid pathway of photosynthesis and its occurrence in dicotyledonous plants.

Johnson, Sr., M.C. (1964). "An electron microscope study of the photosynthetic apparatus in plants with special reference to Gramineae." Ph.D. thesis, University of Texas.

Kiesselbach, T. A. (1916). *Bull. Neb. agric. Exp. Stn.* **6**, 1–214. Transpiration as a factor in crop production.

Kisaki, T. and Tolbert, N. E. (1969). *Pl. Physiol.* **44**, 242–250. Glycolate and glyoxylate metabolism by isolated peroxisomes or chloroplasts.

Kortschak, H. P., Hartt, C. E. and Burr, G. O. (1965). *Pl. Physiol.* **40**, 209–213. Carbon dioxide fixation in sugarcane leaves.

Krenzer, E. G., Jr. and Moss, D. N. (1969). *Crop. Sci.* **9**, 619–621. Carbon dioxide compensation in grasses.

Laetsch, W. M. (1968). *Am. J. Bot.* **55**, 875–883. Chloroplast specialization in dicotyledons possessing the C_4 dicarboxylic acid pathway of photosynthetic CO_2 fixation.

Laetsch, W. M. (1969). *Sci. Prog.* **57**, 323–351. Relationship between chloroplast structure and photosynthetic carbon-fixation pathways.

Laetsch, W. M., Stetler, D. A. and Vlitos, A. J. (1966). *Z. Pflanzenphysiol.* B. **54**, 472–474. The ultrastructure of sugarcane chloroplasts.

Larcher, W. (1969). *Photosynthetica* **3**, 167–198. The effect of environmental and physiological variables on the carbon dioxide gas exchange of trees.

McAlister, E. D. and Myers, J. (1940). *Smithson. misc. Collns* **99**, 6–26. The time course of photosynthesis and fluorescence observed simultaneously.

Meidner, H. (1962). *J. exp. Bot.* **13**, 284–293. The minimum intercellular-space CO_2-concentration of maize leaves and its influence on stomatal movements.

Miller, E. S. and Burr, G. O. (1935). *Pl. Physiol.* **10**, 93–114. Carbon dioxide balance at high light intensities.

Miller, V. J. (1960). *Am. Soc. Hort. Sci.* **75**, 700–703. Temperature effect on the rate of the apparent photosynthesis of seaside bent and bermudagrass.

Mollenhauer, H. H., Morré, D. J. and Kelly, A. G. (1966). *Protoplasma* **62**, 44–52. The widespread occurrence of plant cytosomes resembling animal microbodies.

Moss, D. N. (1962). *Nature* **193**, 587. The limiting carbon dioxide concentration for photosynthesis.

Moss, D. N. (1967). *Pl. Physiol.* **42**, 1463–1464. High activity of the glycolic acid oxidase system in tobacco leaves.

Moss, D. N. (1968). *Crop Sci.* **8**, 71–76. Photorespiration and glycolate metabolism in tobacco leaves.

Moss, D. N., Musgrave, R. B. and Lemon, E. R (1961). *Crop Sci.* **1**, 83–87. Photosynthesis under field conditions. III. Some effects of light, carbon dioxide, temperature, and soil moisture on photosynthesis, respiration, and transpiration of corn.

Moss, D. N., Krenzer, E. G., Jr. and Brun, W. A. (1969). *Science* **164**, 187–188. Carbon dioxide compensation points in related plant species.

Moss, D. N. and Rasmussen, H. P. (1969). *Pl. Physiol.* **44**, 1063–1068. Cellular localization of CO_2 fixation and translocation of metabolites.

Murata, Y. and Iyama, J. (1963). *Proc. Crop Sci. Soc. Japan* **31**, 315–322. Studies on the photosynthesis of forage crops. II. Influence of air temperature upon the photosynthesis of some forage and grain crops.

Murata, Y., Iyama, J. and Honma, T. (1965). *Proc. Crop Sci. Soc. Japan* **34**, 154–

158. Studies on the photosynthesis of forage crops. IV. Influence of air temperature upon photosynthesis and respiration of alfalfa and several southern type forage crops.

Osmond, C. B. (1967). *Biochim. biophys. Acta* **141**, 197–199. β-Carboxylation during photosynthesis in *Atriplex*.

Osmond, C. B., Troughton, J. H. and Goodchild, D. J. (1969). *Z. Pflanzenphysiol.* **61**, 218–237. Physiological biochemical and structural studies of photosynthesis and photorespiration in two species of *Atriplex*.

Prat, H. (1936). *Ann. Sci. Nat. Bot.*, sér 10, **18**, 165–258. La Systématique des Graminées.

Rabinowitch, E. I. (1951). "Photosynthesis and Related Processes", Vol. II, Part I. Interscience Publishers, New York.

Racker, E. (1957). *Archs Biochem. Biophys.* **69**, 300–310. The reductive pentose phosphate cycle. I. Phosphoribulokinase and ribulose diphosphate carboxylase.

Rhoades, M. M. and Carvalho, A. (1944). *Bull. Torrey bot. Club* **71**, 335–346. The function and structure of the parenchyma sheath plastids of the maize leaf.

Rosado-Alberio, J., Weier, T. E. and Stocking, C. R. (1968). *Pl. Physiol.* **43**, 1325–1331. Continuity of the chloroplasts membrane systems in *Zea mays* L.

Shantz, H. L. and Piemeisel, L. N. (1927). *J. agric. Res.* **34**, 1093–1189. The water requirements of plants at Akron, Colorado.

Slack, C. R. (1969). *Phytochem.* **8**, 1387–1391. Localization of certain photosynthetic enzymes in mesophyll and parenchyma sheath chloroplasts of maize and *Amaranthus palmeri*.

Slack, C. R. and Hatch, M. D. (1967). *Biochem. J.* **103**, 660–665. Comparative studies on the activity of carboxylases and other enzymes in relation to the new pathway of photosynthetic carbon dioxide fixation in tropical grasses.

Slack, C. R., Hatch, M. D. and Goodchild, D. J. (1969). *Biochem. J.* **114**, 489–498. Distribution of enzymes in mesophyll and parenchyma—sheath chloroplasts of maize leaves in relation to the C_4-dicarboxylic acid pathway of photosynthesis.

Smalley, A. E. (1959). *Proc. Salt Marsh Conf. Sapelo Island, Georgia.* "The growth cycle of *Spartina* and its relations to the insect populations in the marsh."

Stebbins, G. L. (1956). *Am. J. Bot.* **43**, 890–905. Cytogenetics and evolution of the grass family.

Thornthwaite, C. W. (1933). *Geogrl. Rev.* **28**, 433–440. The climates of the earth.

Tolbert, N. E., Oeser, A., Hageman, R. H. and Yamazaki, R. K. (1968). *J. biol. Chem.* **243**, 5179–5184. Peroxisomes from spinach leaves containing enzymes related to glycolate metabolism.

Tolbert, N. E., Oeser, A., Yamazaki, R. K., Hageman, R. H. and Kisaki, T. (1969). *Pl. Physiol.* **44**, 135–147. A survey of plants for leaf peroxisomes.

Tranquillini, W. (1964). *A. Rev. Pl. Physiol.* **15**, 345–362. The physiology of plants at high altitudes.

Tregunna, E. B. (1966). *Science* **151**, 1239–1241. Flavin mononucleotide control of glycolic acid oxidase and photorespiration in corn leaves.

Tregunna, E. B. and Downton, J. (1967). *Can. J. Bot.* **45**, 2385–2387. Carbon dioxide compensation in members of the Amaranthaceae and some related families.

Tregunna, E. B., Krotkov, G. and Nelson, C. D. (1961). *Can. J. Bot.* **39**, 1045–

1056. Evolution of carbon dioxide by tobacco leaves during the dark period following illumination with different light intensities.

Tregunna, E. B., Krotkov, G. and Nelson, C. D. (1964). *Can. J. Bot.* **42**, 989–997. Further evidence on the effects of light on respiration during photosynthesis.

Warburg, O. (1920). *Biochem. Z.* **103**, 188–217. Uber die Geschwindigkeit der photochemischem Kohlensaurezersetzung in lebender Zelien. II.

Zelitch, I. (1966). *Pl. Physiol.* **41**, 1623–1631. Increased rate of net photosynthetic carbon dioxide uptake caused by the inhibition of glycolate oxidase.

Zelitch, I. (1967). *In* "Harvesting the Sun". (San Pietro, A., Greer, F. A. and Army, T. J., eds.), pp. 231–248. Academic Press, New York. Water and CO_2 transport in the photosynthetic process.

Zelitch, I. (1968). *Pl. Physiol.* **43**, 1829–1837. Investigations on photorespiration with a sensitive ^{14}C-assay.

Zelitch, I. and Day, P. R. (1968). *Pl. Physiol.* **43**, 1838–1844. Variation in photorespiration. The effect of genetic differences in photorespiration on net photosynthesis in tobacco.

Ecological Aspects of Fishery Research

J. A. GULLAND[1]

Fishery Economics and Institutions Division, Food and Agriculture Organization, Rome

I. Introduction

The ecology of the seashore has always taken its full share of general ecological studies—university professors are as glad as anyone else to spend part of the summer by the seaside. However, the ecology of the open sea, of which various aspects of fishery research form a major part, has to a large extent become separated from the general field of ecological studies.

Some of this separation is due to the difficulty and expense of any

[1] The views expressed here are those of the author and not necessarily those of F.A.O.

ocean research—ocean science has been compared to studying the earth by means of instruments hanging from a balloon a thousand feet up in thick fog—and the daily costs of such an operation, for even the smallest research vessel capable of staying at sea in all weather conditions, runs into hundreds of dollars. Hence marine scientists have concentrated on matters likely to produce results of foreseeable practical value.

However, the fishery scientist does have some advantages. He is working in an environment which is far less changed from its natural state than the land. Also the activities of the ordinary fishermen can provide him with very extensive information about those natural populations which are of commercial interest. Thus available statistics provide information on the time and place of capture, and the size and sex (and also size of the foetus of pregnant females) of virtually all whales killed in the past fifty years, and these comprise the majority of all deaths of large whales in this period. Also there is often information on the fishing effort—the number of days whale-catchers have been hunting for whales, or the number of hours that trawlers have had their nets on the bottom—which, combined with data on the numbers caught, can provide useful estimates of the abundance, at least in relative terms.

Also, the applications of fishery research differ from those of most terrestrial ecology. On land, man is largely engaged in maintaining large populations of desired and protected species (his crops and herds of animals) under more or less unnatural conditions, and when engaged in destroying individuals of natural populations the aim is usually to kill pests as cheaply and effectively as possible—preferably with few or no survivors. In the sea the species that man expends most time and cost in catching and killing are precisely those species—the fish of commercial importance—which he is most interested in maintaining at a high level of abundance. Crudely, therefore, an applied terrestrial ecologist might be prepared to see his problems in qualitative terms—good for the crops, or good for the pests—while a fishery scientist must from the start become involved in quantitative expressions which are at least mathematically more sophisticated.

This difference is perhaps most obvious in the use of the term density dependence. Terrestrial ecologists have discussed at length definitions of density dependence, and its relations to the problem of control of population abundance (e.g. Andrewartha and Birch, 1954; Nicholson, 1954, 1958), while fishery scientists have been more concerned with the quantitative mathematical form of the average relation between density and various population parameters, especially reproductive rate (e.g. Ricker, 1954).

There are other big differences; for example, the ocean scientist has to deal with plants whose size is measured in fractions of a millimetre, and lifespan in days, if not hours, rather than months and years, and therefore the ecological aspects of fishery research are far from being a wet and rather blurred copy of terrestrial ecology. Despite the difficulties, there are some directions, especially in the quantitative aspects, for example population dynamics, in which marine scientists have advanced at least as far as their terrestrial counterparts.

Some of these advances and some other aspects of fishery research likely to be of interest to the general ecologist are discussed below, although it is not easy in one paper to present a complete or balanced account of all aspects of present-day fishery research.

II. The Application of Fishery Research

Most fishery research is carried out in government laboratories, so that the research is expected to produce results of clear practical value. That is, it should be research with applications, but more than applied research, since fundamentally new ideas have to be developed.

A fishery research worker must be able to see how the results of his work may be applied to the interests of the fishing industry (in its widest sense), but to achieve this aim his field of study must include research in marine biology, physical oceanography, etc. The applications can most easily be discussed according to the timing of their effect on the fisheries—in the next year or so, or in the next decade.

A. Immediate Improvement of Catches

The immediate effects are concerned mainly with direct improvement of catches. In this there have recently been great advances, although the biggest advances have been in various technological aspects rather than through the application of biological sciences. Fishing may still, in some senses, be in the hunting or primitive pre-agricultural phase, but it is hunting with very advanced weapons. The modern fishing vessel has a wealth of modern equipment for detecting, catching or processing the fish; the largest Russian factory ship now being completed is as large as an ocean liner (40 000 gross tons), and many vessels have more numerous and more complex electronic equipment than warships of only twenty years ago. In this the ecologist cannot contribute much, although it is clear that the most efficient catching techniques can be evolved only when the engineer designing new equipment has close links with biologists familiar with the behaviour of the fish, especially their reactions to moving ropes and netting, etc. Ecologists can give guidance on the most promising areas for fishing; in this sense any good fisherman is a practical ecologist. A trawlerman, for instance, knows quite well the connection of various demersal fishes with each other and with particular types of bottom and bottom fauna, and in established fisheries the normal pattern of distribution of the fish in space and time is well known.

Deviations from this pattern will often occur, and predictions, or rapid information concerning these, can greatly help the fisherman. These deviations can be related to changes in the ocean climate and weather—the temperature of the water, the strength or position of currents, etc.—and as the general understanding of these improves, so will improve the ability of forecasting, just as weather forecasting on land is improving.

B. FORECASTING

In the slightly longer term the fish industry requires information on catches for the next year or two ahead. In established fisheries such forecasts are often less useful for the fisherman than for the industry ashore which is concerned with processing, distribution and marketing. Changes in total landings are often unpredictable and in most fisheries, especially those dealing with fish for direct marketing for human consumption, are usually followed by rapid and sometimes violent changes in prices. These price fluctuations tend to leave the total return to the individual fisherman fairly constant or even to make it fall when there is a glut of fish on the market. Thus the critical factor in a fisherman's income is not the average level of catches but how his catches compare with those of other fishermen landing at the same time. A wholesale merchant or processor depends more on achieving a good balance between the available supplies and the facilities for selling and processing, for which good forecasting of total future landings is most valuable.

The forecasting of catches of most temperate and boreal species of fish for one or two years ahead is in principle quite easy. The fish are long lived, and there is often a great variability in year-class strength (the numbers of fish born in any year surviving to a catchable size). Thus it is possible at the time of writing (in middle of 1968) to forecast that the fishery for cod (*Gadus morhua*) round Bear Island in 1972 will be poor, because all the fish that will contribute to this fishery are already alive, and it is known that the year-classes principally involved —those between 1963 and 1968—are all below, or well below, average strength.

The effect of these year-class fluctuations on some of the world's major sea fisheries has been known for a long time, the classic example being the Norwegian herring (*Clupea harengus*) for which the outstanding 1904 year-class provided the bulk of the catches from 1910 until

1914, and was still the dominant year-class in 1920 (Hjort, 1914, 1926). For some time forecasting was based mainly on data obtained directly from the fishery, since a proportion of a year-class may enter the fishery a year or two before the year-class as a whole makes its biggest contribution to the catches (see Fig. 1, adapted from Anon., 1966). The

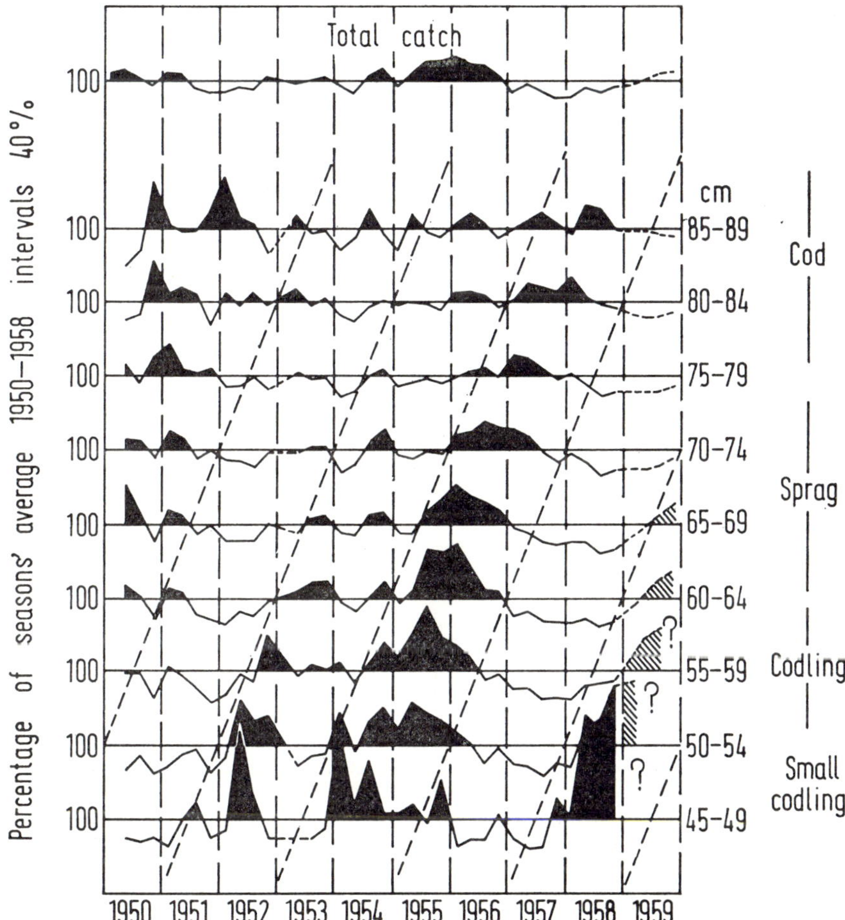

FIG. 1. Forecasting of catches of cod from the Bear Island region, showing the progression of strong year-classes through the length groups (from Anon., 1966).

strength of year-classes can also be measured before the fish enter the fishery. Figure 2 shows the correlation derived from the work of M. Grosslein of Woods Hole, between the index of abundance of young haddock (*Melogrammus aeglefinus*) on Georges Bank, as measured by research vessels in the autumn of the year of birth, and the catches per

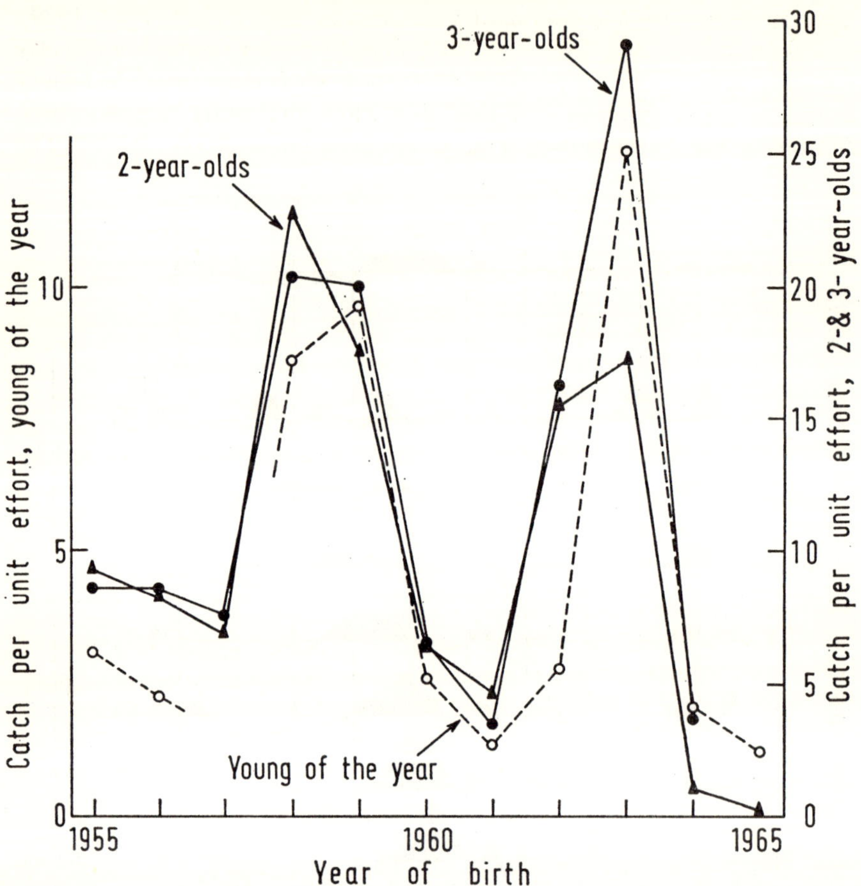

Fig. 2. Measures of abundance of year-classes of Georges Bank haddock, showing the close agreement between estimates from research vessel samples when a few months old (broken line), and in the commercial fishery when 2 and 3 years old (full lines).

unit effort in the commercial fishery when 2 and 3 years old (Graham, 1966). The most extensive data of this kind on the distribution of pre-recruit fish have been collected by U.S.S.R. surveys in the Barents Sea (e.g. Baranenkova, 1960). These have been carried out by research trawlers, using fine-meshed nets to retain the small fish and have involved some hundred hauls per year. This work has been intensified, and, as a co-operative effort involving also British and Norwegian research vessels and advanced acoustic methods, now enables estimates to be made of the abundance of young fish in their first year of life, for herring, haddock and others as well as cod (Anon., 1966).

These forecasts might be made earlier, and without the need for expensive surveys of pre-recruit fish, if there were a better understanding of the factors controlling year-class strength. It may be deduced from the fact that year-class can be measured fairly accurately early in life (e.g. at about 9 months old in the Arcto-Norwegian cod, which does not reach commercial size until 3–5 years old), that the critical factors operate at least before the time at which these estimates are made. Many attempts have been made to correlate year-class strength with various environmental factors, but the results have often been inconclusive, and few have stood the test of time (e.g. Carruthers, 1938; Gulland, 1953; Saville, 1959; 1965, for the North Sea haddock; Corlett, 1965, and Gulland, 1965, for the Arcto-Norwegian cod). Many factors have been considered—wind strength and direction (Carruthers *et al.*, 1951), temperature (this seems to give a good relationship for the West Greenland stock of cod, which is close to the limit of its distribution, Herman *et al.*, 1965), rainfall (for the tropical shrimps of Australia and the Gulf of Mexico which pass some of their early stages in brackish waters)—although it can be argued on theoretical grounds that the factor most likely to be significant is the abundance of food.

C. CONSERVATION AND MANAGEMENT

In the long term the biggest impact of the ecological sciences on the fishing industry is produced by conservation and management of fishery resources. The scientific aspects of these problems are considerable, and are discussed in detail in later sections of this paper. However, so far as the impact on the fishing industry is concerned, the scientific conclusions, or at least their most important aspects, can be summarized in a single curve relating the amount of fishing to the average catch. Two such curves are shown in Fig. 3; these correspond to two of the simpler and best known models, that of Schaefer (1954), and the constant parameter curve of Beverton and Holt (1957). While these curves differ at very high levels of fishing—the Schaefer model predicts extinctions of the population, the other no change in catch, and this difference may become critical at some stage in the management of a fishery—the difference is less important than the similarities at lower levels of fishing.

The relevance of these curves to the economics of the fishery, and their similarities, are best shown in terms of the marginal yield, i.e. the increase in total catch resulting from a unit increase in the amount of fishing (Larrañeta, 1967; Gulland, 1968a). The marginal yields corresponding to the curves in Fig. 3 are given as the full lines in Fig. 4. Both curves decrease steadily with increasing fishing, the marginal yield becoming small at levels of fishing which are not very high, one curve in fact soon becoming negative. Thus even at moderate levels the additional costs of more fishing will be much more than the value of the

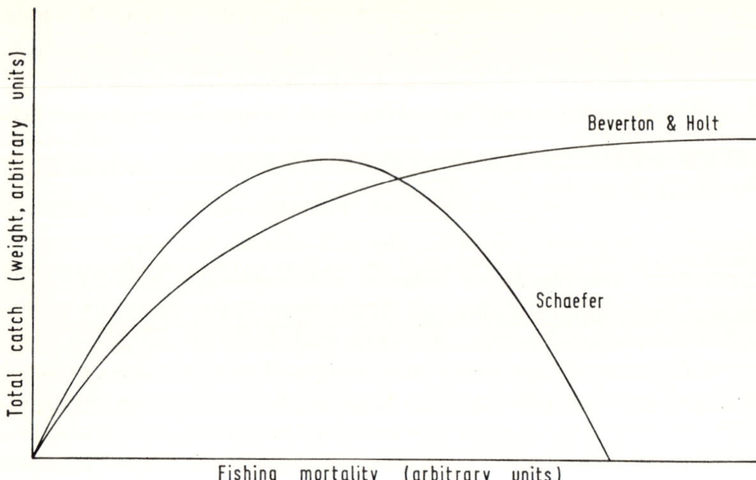

FIG. 3. The relation between the amount of fishing and the total catch, as predicted by two of the most widely used models, that of Schaefer (1954) and the constant parameter model of Beverton and Holt (1957).

increase in the total catch (and indeed in extreme cases the catch may decrease). If a fishery operated as a single unit, economic factors would prevent too great expansion and stabilize the fishing effort at around the level at which the value of the marginal yield is equal to the cost of the additional effort. There would then be no great urgency for scientific analysis of the effect of fishing on the stocks.

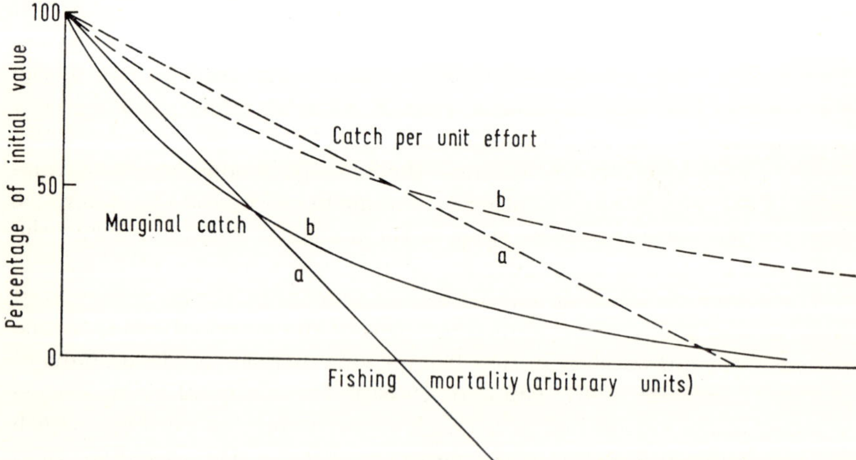

FIG. 4. The relation between the amount of fishing and the marginal catch (the increment to the total catch by one additional unit of effort—full lines), and the catch per unit effort (broken curve), as predicted by the two models illustrated in Fig. 3—(*a*) Schaefer, and (*b*) Beverton and Holt.

The economic factor that determines whether a fisherman buys a new and more powerful boat, or a company invests money in increasing its fleet, is not the contribution the extra fishing makes to the total catch but the actual catches taken by the new vessels. These catches will be proportional to the catch per unit effort in the fishery, which is equal to the slope of the line joining the origin to the appropriate point in the curves of Fig. 3. The relation between catch per unit effort and total effort is given by the broken lines in Fig. 4. Except at very low levels of fishing the catch per unit effort is more, and often very much more, than the marginal yield, so that the amount of fishing may well increase beyond the optimum economic level or even the level giving the maximum catch. Even when the fishermen realize that too much fishing is being done, no improvement is possible unless action to limit the amount of fishing is taken more or less unanimously—if only some of those fishing a stock restrict their activities, most of the benefit will go to those who continue fishing without restriction.

Outside national fishery limits the necessary action has to be taken internationally and, as fishing has expanded in different parts of the world, an increasing number of international commissions and similar bodies have been set up to deal with the twin problems of determining, from the best available scientific evidence, the measures (if any) needed to ensure rational exploitation of the resource, and taking administrative measures to introduce and enforce these measures. A list of these bodies is given by Gulland and Carroz (1968).

Since action has to be taken virtually unanimously, the scientific advice needs to be clear and persuasive and must be as accurate as possible. Unless the theoretical basis and the necessary observations and data have been sufficiently advanced, agreement on the facts, and on the necessary action, may come too late. The outstanding example is in Antarctic whaling. In succession the catches of each of the major species (blue (*Balaenoptera musculus*), fin (*B. physalis*) and sei whales (*B. borealis*)) have reached a peak, and then collapsed (see Fig. 5). The catches of fin whales were roughly constant from 1948 to 1962, while the International Whaling Commission set a total catch quota of 14 000 to 15 000 Blue Whale Units (one blue whale = 2 fin whales = 6 sei whales).

By the early 1950's it was becoming clear both to scientists and to the whaling industry that this quota was too high. However, the switching of the main attention of the whaling fleets from the rapidly declining blue whale stocks to the fin whales made it difficult to establish beyond argument the rate (if any) at which the fin whales were decreasing. Thus no clearly agreed advice was given to the industry on the desirable level of quota, until the Special Committee of Three (later

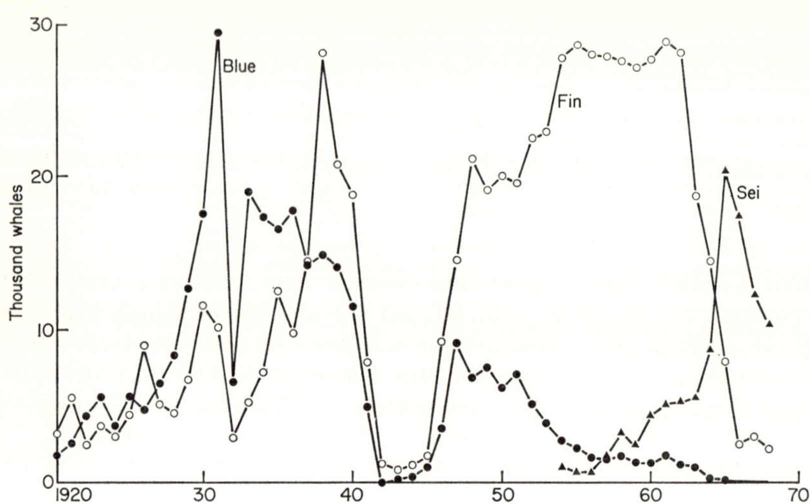

Fig. 5. Total catches of blue, fin and sei whales from the Antarctic, 1919–68.

Four) Scientists was set up by the Commission, and reported in 1963 (Chapman, 1964). By this time the stocks had decreased so much that the Commission was given the unpleasant choice between an immediate reduction of catches to an extremely low level for five to ten years (followed by better catches), or a continuation at around the previous level, with collapse of the stock and the industry in a few years time (with recovery taking many years). Not surprisingly, the second alternative was chosen, with the implicit hope that the scientists were wrong. As Fig. 5 shows, catches of fin whales have indeed collapsed along with most of the important Antarctic whaling industry.

Despite this major failure, the I.W.C. did put some restraint on the degree of exploitation and the value of the increased catches thus achieved has been estimated as £50 million (Gulland, 1966). This is a large sum in comparison with the expenses of the Commission and of the research associated with it. The benefit that might have been achieved if better scientific advice had been available earlier, and had been acted on, would have been very much greater. These figures give some measure of the potential benefits from rational management, if the necessary scientific and administrative problems can be solved.

III. Models of the Dynamics of Fish Populations

A. Types of Model

Several models have been used to represent the dynamics of fish populations, and especially their reaction to fishing, and they can be conveniently grouped in two classes (Schaefer and Beverton, 1963). One treats the population as a single unit, subject to simple laws of population growth (Volterra, 1931; Leslie, 1957), on which the fishery acts as a predator in a simple predator/prey system (e.g. Graham, 1939; Schaefer, 1954). The other considers the population as the sum of its individuals, and the abundance of the population is determined from the numbers of young entering the population per unit time, their growth and their mortality. Fishing acts as an additional mortality (e.g. Baranov, 1918; Ricker, 1948, 1958; Beverton and Holt, 1957).

These two types may be considered less as alternatives than as being applicable at different stages of research. At first, when information is scanty, the Schaefer type of model gives a convenient and fairly realistic description of how a population is affected by fishing. Few data (essentially only on population abundance and the amount of fishing) are required to apply the model. However, it is difficult to improve the fit of the model to an actual situation by using further information, e.g. on the growth of the individual fish, as this becomes available. Also, the model gives only a limited insight into the factors controlling the population. The other type of model requires more data, and in its simpler forms, in which the parameters of recruitment, growth and natural mortality are taken as constant, appears less realistic. However, it is relatively easy, at least in principle, to include in the model variations in these parameters.

B. THE UNIT STOCK

Before applying the models to the situation in a particular fishery, it is necessary to define the population, or stock, of fish concerned. Fishery research is usually concerned less with the individual fish or the total population of the species, than with the group or stock of fish exploited by the particular fishery of immediate interest. For instance, the cod lives, and is fished, right across the North Atlantic from New England to the north Russian coast, but the stock of cod off Labrador differs very greatly from the stock in the North Sea, both in biological characteristics—growth, natural mortality, etc.—and in the history of the fishery on it. The two stocks must, therefore, be studied separately.

Ideally, a unit stock is a self-contained and self-perpetuating group, with no mixture from outside, and within which the biological characteristics and the impact of fishing are uniform. Such a unit stock would also be a genetic unit.

The identification and definition of a unit stock will often be in terms of spawning, especially for those fish which have long dispersions or migrations over a wide area outside the spawning time but return to spawn at a single restricted spawning ground. Thus, the Arcto-Norwegian cod living in the area from north Norway and Russia to Spitsbergen form a clear unit stock. The juveniles and the adults in the summer are spread over most of the ice-free part of the Barents Sea shelf, but in early spring the adults collect for spawning in a small area inside the Lofoten Islands (Cushing, 1966a). Thus there is a high degree of mixing among all the adults in the stock. Also, though some few individuals may move from the north-east Arctic to the Faroes (Holden, 1960) or elsewhere, the degree of mixing between the Arctic cod and these other stocks appears to be too small to have a significant effect on the analyses described in this paper which are mainly concerned with events within one generation. However, the mixing may be great enough to prevent any genetic distinction.

Few groups of fish form such neat units, and the choice of what should be considered a unit stock has usually to be made empirically, in the light of the extent of the data available and the precision and detail required in the analysis. The actual choice is likely to change as more information becomes available, or the objectives of the analysis change. Thus, for example, when considering factors in the ocean affecting the Pacific salmon, including the coastal fishing of the returning adults, all the fish coming from a particular river system may be considered as a unit stock, e.g. the Fraser River sockeye. For detailed studies of the early history it may, however, be necessary to consider each group of fish spawning in a particular stream as a unit stock.

In practice, it is usually convenient to treat as large a group as possible as a unit stock, thus enlarging the amount of information available for analysis, and avoiding the need to subdivide the statistical information on catches, etc. Even when the group of fish being treated as a unit in fact contains several independent unit stocks, serious errors will not be incurred if the values of the parameters of growth, mortality, etc., do not vary much between the stocks. If all the fish grow at the same rate, then clearly any measurement of growth will give the same result whatever the source of the samples. Of the various parameters, the fishing mortality is the most likely to be very different between adjoining stocks, especially when the fleets concerned consist of small, short-range vessels. Exceptionally, fishery differences may make it necessary to consider separately different parts of the same unit stock. Thus there is little mixing between the two immature groups of Arcto-Norwegian cod, living in the Bear Island-Spitsbergen area, and the eastern Barents Sea, until they come together for spawning. During the five to seven years between reaching a fishable size and becoming mature, the abundance of a year-class will be greatly affected by fishing. In general, the fishing mortality is different in the two areas, and there are also some differences between them in growth and in the relative strength of different year-classes.

Conversely, if fishing is spread fairly uniformly over a wide area, then the separation of different stocks (if any) within the area is less important. Thus, the Peruvian anchovy fishery extends some hundreds of miles; it is unlikely that there is complete mixing of fish along this length. However, the fishery is continuous along the coast, and movements of the fleet up and down it tend to make the fishing mortality reasonably uniform. Thus, analyses of the state of the stocks treating the catches in the north and south separately have given the same conclusion, and the same as that obtained from treating the fishery as a single whole (Schaefer, 1967b).

The problem of the definition of the unit stock is one common to all studies of natural populations, if most prominent in fishery research. Some of the apparent inconsistencies in studies of the population dynamics of other animals seem to have arisen because the group of animals being studied did not form, to a sufficiently close approximation, a unit stock.

C. RECRUITMENT TO THE EXPLOITED STOCK

In both models it is usual to consider in detail only a segment of the total population—the exploited or recruited stock. Thus the early life stages—eggs, larvae, etc. are normally excluded from most of the analyses. The primary reason for this is simply the practical one that the extensive data available from the commercial fisheries give no information on these stages (with some exceptions such as the whitebait fishery for small sprats (*Clupea sprattus*) and herring, or the Japanese "schivashu" fishery (Nakai *et al.*, 1955)). There are also good scientific reasons. After spawning there is very little interaction between adults and young, which occupy quite different positions in the ecological system. A mature cod is a large and active predator on other fish, crabs, etc., but a newly hatched cod larva is a member of the planktonic community, preyed on by small fish, ctenophores, etc.

A further distinction may be made between the exploited stock—the group of fish from which the catches actually come—and the recruited, or exploitable stock, which will include fish which could be caught, but which are in fact not liable to be caught with the particular gear in use. An obvious example is provided by the small fish which can pass through suitably sized meshes of a trawl. In some fisheries young and old fish may be present in the same area and the choice of the dividing line is somewhat arbitrary. In others there is a clear change in habit, e.g. the cod (*Gadus morhua* L.) in the Barents Sea is planktonic or pelagic until

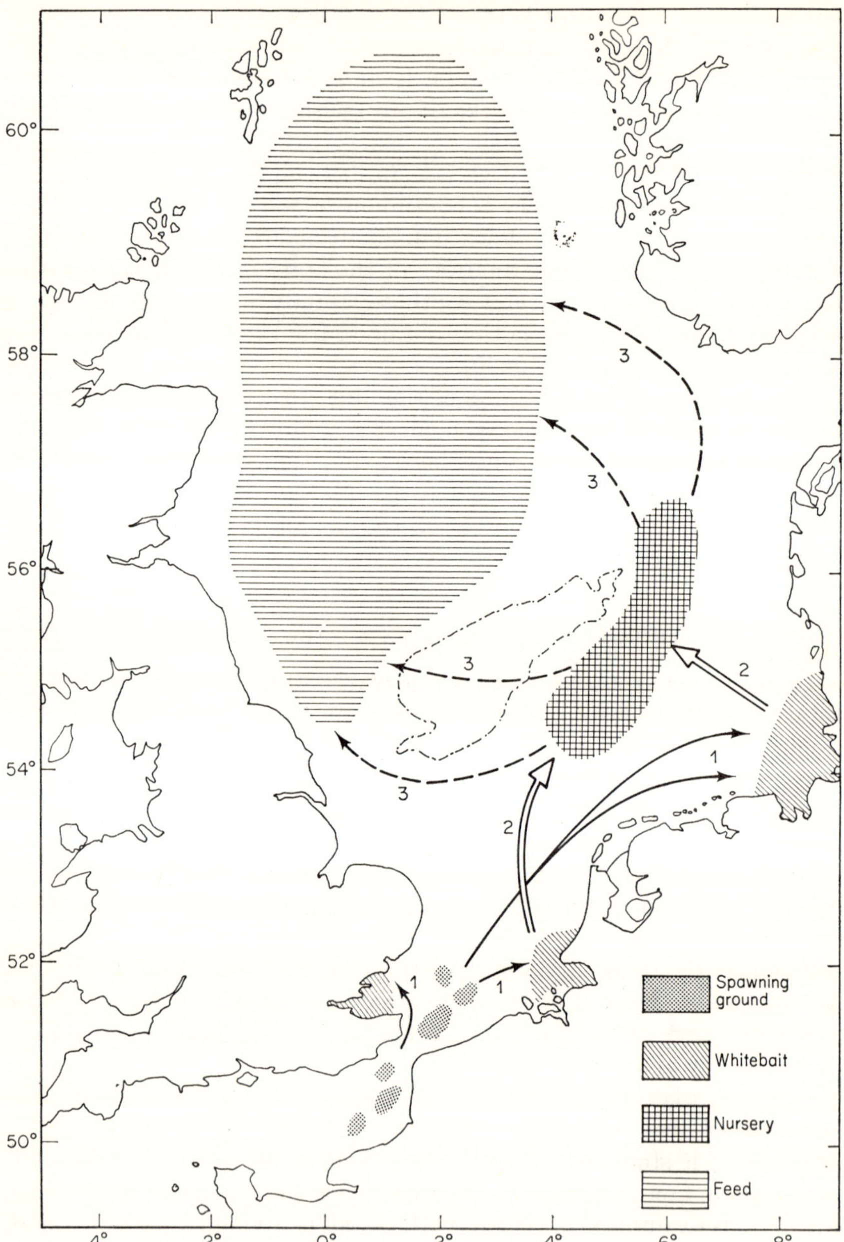

Fig. 6. Recruitment and migration patterns of North Sea herring: (1) from the spawning grounds to the inshore nursery where they are occasionally fished as whitebait; (2) on to the juvenile, Bloden ground; (3) on to the main fishing grounds, and into the adult stock (derived from Cushing, 1968).

the autumn of its first year of life, when it takes to the bottom (at around 10 cm long), and becomes vulnerable to trawling. The exploitable phase starts therefore at 10 cm, but the cod are not intensively exploited until they reach about 40 cm (ICES, 1966a).

For other fish, recruitment may involve a definite movement. Figure 6, using data from Cushing (1968), shows the recruitment pattern of the Downs group of North Sea herring. This consists of three stages: (1) a drift from the spawning grounds to the inshore ("whitebait") grounds; (2) an offshore movement when about 1 year old to the nursery areas of the Bloden ground (Bertelsen and Popp Madsen, 1953); and (3) a movement from these areas, some two years later, to the feeding areas of the adult stock.

Until recently in the North Sea only the adult herring were fished to any great extent (the whitebait fishery is extremely small), and recruitment to the exploited stock was effected by the movement from the nursery area to the feeding area. Since 1951 an important fishery for young herring has developed on the Bloden ground (ICES, 1966b). For any analysis that takes this fishery into account the exploited stock should therefore be considered as including the fish on the Bloden ground as well as the adult stock.

Whatever the recruitment patterns of the fish stock concerned, nearly every analysis of its population dynamics, especially of the effects of fishing, makes a distinction between the exploited stock and the rest of the fish stock even though this may have to be somewhat arbitrary. For the following sections the term stock will be taken as meaning the exploited stock unless explicitly stated otherwise.

D. Dynamic Pool Models

In these models the stock is presumed to be adequately represented by a single quantity, generally the population biomass, B, or sometimes (e.g. whales) by the population numbers (Chapman, 1964). It is then assumed that, in the absence of exploitation, the population has a natural rate of increase which is a function of the population, i.e.

$$dB/dt = f(B) \tag{1}$$

where $f(B) = 0$ for $B = 0$ and $B = B_{max}$, the maximum population.

If the rate of removals by the fishery is equal to $f(B)$, then the population will remain constant. $f(B)$ may then be referred to as the equilibrium catch (Watt, 1968) or sustainable yield (Schaefer, 1954), and we can write, for the equilibrium catch in unit time

$$f(B) = FB \tag{2}$$

where F = fishing mortality coefficient

or
$$f(B) = qfB \tag{3}$$

where q = catchability coefficient
f = fishing intensity, or fishing effort.

It may be noted that the relation $F = qf$ can be regarded as a definition of the fishing effort rather than as an assumption that needs to be established. The assumption which has to be made in any specific fishery is that the statistics available—number of nets used, or of hours fishing—are in fact adequate measures of the actual fishing effort. A number of studies have been made of ways in which this assumption can fail, e.g. changes in the size and power of trawlers, reduction of the effective effort by lines as the hooks become occupied (Margetts, 1949a, b; Gulland, 1955, 1956; Beverton and Holt, 1957). Satisfactory measures (e.g. the time the trawl is on the bottom × size of trawler) are available for trawls, but measurement of effort in other fisheries, especially those in which much of the time is spent searching for fish, rather than actively fishing, is one of the many outstanding problems in fishery research (Gulland, 1964, 1968c).

If the simple law of logistic growth is assumed, then it follows that

$$f(B) = aB(B - B_{max}) \tag{4}$$

which gives a parabolic relation between equilibrium catch and population biomass (Schaefer, 1954, 1957). Further, from (3) it follows that

$$a(B - B_{max}) = qf \tag{5}$$

that is, there is a linear relation between population abundance (which, with a suitable measure of effort, will be proportional to the catch per unit effort) and fishing effort.

Where the range of values of fishing efforts experienced is not too large, a straight line will usually describe adequately the relation between effort and catch per unit effort, e.g. in the fishery for yellowfin tuna (*Thunnus albacares* (Bonnaterre)) in the eastern tropical Pacific (Fig. 7, from Schaefer, 1967a). In this figure the open circles are the years up to 1955, covered by Schaefer's early (1957) analysis. The later years, indicated by closed circles, do not depart very greatly from the line through the early points, but the least squares fit to the post-1956 points (broken line) clearly has a smaller slope, suggesting a curvilinear relation. Such a relation is clear for the stocks of plaice (*Pleuronectes platessa*) and haddock at Iceland (Fig. 8, from Gulland, 1961). Curves relating the

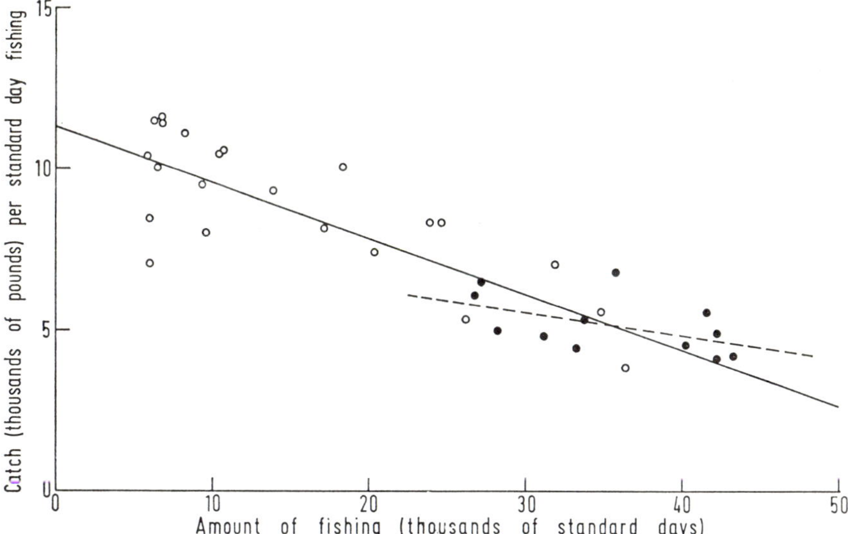

FIG. 7. Relation between the amount of fishing on the yellowfin tuna in the eastern tropical Pacific, and the catches per unit effort, distinguishing the period up to 1955 (open circles) and since (closed circles) (from Schaefer, 1967a).

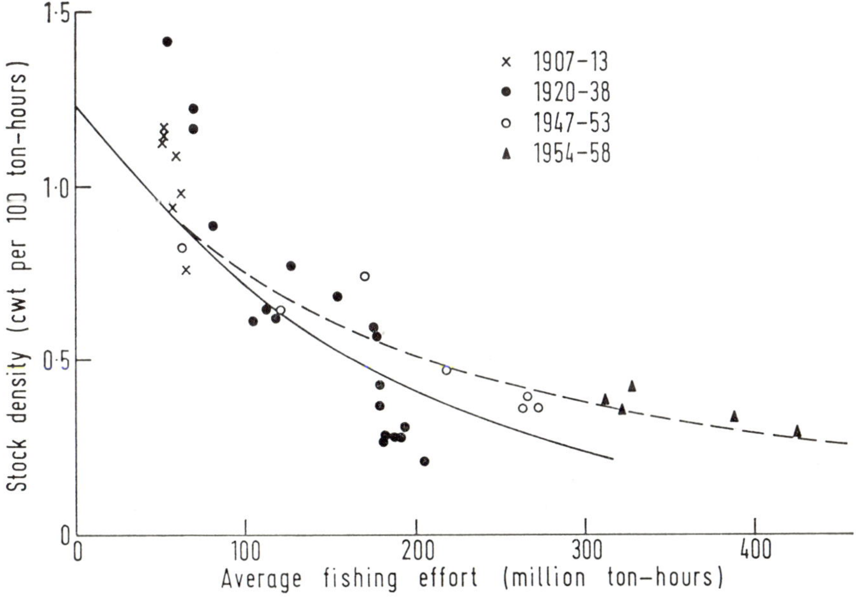

FIG. 8. Relation between the amount of fishing on the haddock stocks at Iceland, and the catch per unit effort of English trawlers. Curves were drawn by eye through the points for the years up to 1953 (full line), and later (broken line) (from Gulland, 1961).

Note. The difference after 1953 can be explained by measures to protect the small fish.

catch per unit effort to effort have been drawn by eye, and from these the equilibrium catch can be deduced as a function of fishing effort (cf. Fig. 3, or Figs 16 and 23 of Gulland, 1961), or a function of the population abundance (catch per unit effort (Fig. 9)). The latter curves depart quite widely from the symmetrical curve—with a maximum at half the maximum population abundance—which is predicted from the logistic model. The actual maxima in Fig. 9 occur at around 30% of the

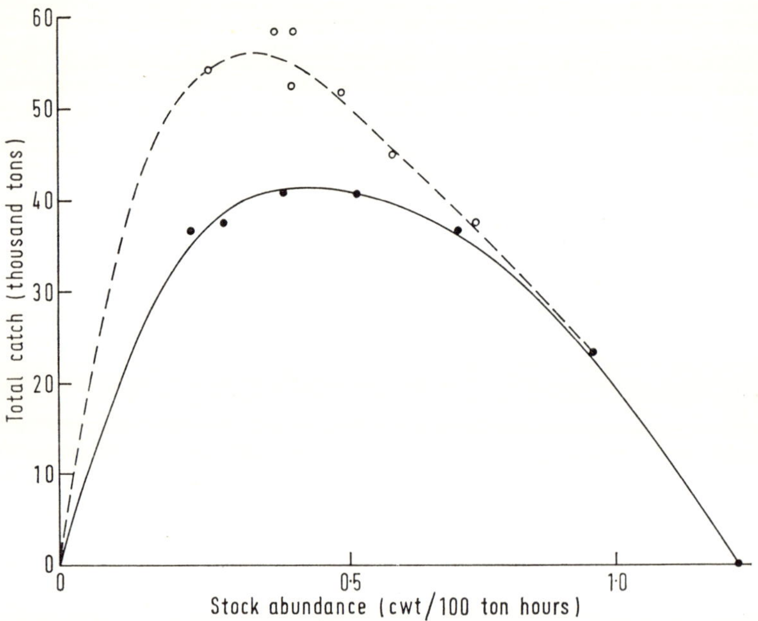

Fig. 9. The average annual catch of Iceland haddock, deduced from the curves in Fig. 8, expressed as functions of the stock abundance.

unfished abundance. It may be noted that the analytic models described in the next section usually also suggest maxima at rather less than half the initial population. In contrast the maxima for the Antarctic blue whale stock appears to be at rather greater than half the unfinished population (Chapman, 1964, Fig. 6). The difference probably lies in the difference between the relation of adult stock to subsequent number of recruits, for whales and for fish. With their high fecundity only a small adult stock of fish may be necessary to ensure adequate recruitment.

This type of model has the advantage of simplicity, and of fitting in with other simple models of predator-prey systems. Schaefer (1954) has emphasized this similarity by describing the development of the fishing effort itself as a function of itself (predator abundance) and the catch per unit effort (prey abundance). If the linear form of equation

(5) does not give an adequate fit to the data, some more suitable function can be deduced (e.g. Pella, 1967).

The probable non-linearity of the relationship between catch per unit effort and effort is not the only or most serious objection to this model. The net change in population abundance is clearly not determined solely by the population abundance, but will depend also on extrinsic factors, and on the composition of the population. A population that has been reduced by fishing to mainly young individuals will increase faster (at least through growth less natural mortality) than will one of the same biomass, but composed of old individuals (recent year-classes having been weak). For many stocks, including some heavily exploited stocks (e.g. haddock in the North Sea) the main cause of the year-to-year changes in abundance is the variation in year-class strength. These variations may have been determined several years previously, and often appear to be only weakly correlated, if at all, with the abundance of the adult stock (Parrish, 1956). Because at a time when strong year-classes are becoming old and the entering year-classes are weak, the stock will decrease whatever catch has been taken, the term "sustainable yield" is a misleading one (Gulland, 1968b), and "equilibrium catch" describes better the ideas involved.

Fitting the model to the data of an actual fishery presents no problem if there have been a number of periods in each of which the fishing has been at a fairly constant level, so that the population, at least towards the end of each period, can be considered as being in equilibrium. The observed average annual catch can then be taken as being equal to the equilibrium catch $f(B)$, and related directly to B, graphically or otherwise. Most fisheries, however, have not been so obliging, and many of the most pressing demands for advice from fishery scientists have come when a fishery has been rapidly expanding, and the population is very far from being in equilibrium.

Schaefer (1957) corrected for changing population abundance by writing

$$f(\bar{B}_i) = C_i + (B_{i+1} - B_i) \tag{6}$$

where B_i, B_{i+1} are the values of the population biomass at the beginning and end of the ith period, during which the catch is C_i, and the mean biomass $\bar{B}_i$.

The biomass at an instant of time is more difficult to measure than the mean biomass over a period, so that, writing the biomass at the beginning of the ith interval as the average of the mean biomass during the $i-1$th and ith intervals,

$$\begin{aligned} f(\bar{B}_i) &= C_i + \tfrac{1}{2}(\bar{B}_i + \bar{B}_{i+1}) - \tfrac{1}{2}(\bar{B}_{i-1} + \bar{B}_i) \\ &= C_i + \tfrac{1}{2}(\bar{B}_{i+1} - \bar{B}_{i-1}) \end{aligned}$$

or, if the catches per unit effort, $\bar{U}_i$ etc. are equal to $q\bar{B}_i$ etc.

$$f(\bar{B}_i) = C_i + \tfrac{1}{2} \cdot \frac{1}{q} \cdot (\bar{U}_{i+1} - \bar{U}_{i-1}). \tag{7}$$

If an estimate of q is available, the equation (7) can be used to obtain pairs of values of $\bar{B}_i$ and $f(\bar{B}_i)$, from which the relations can be determined. However, if q can be determined, then it will usually be possible to use the analytic models of the next section (for instance, one of the difficult parameters to estimate is the fishing mortality, F, but this is given at once by the relation $F = qf$).

Equation (7), as an approximation to the equilibrium conditions, ignores the fact that the net natural increase will depend on events in previous seasons as well as on the actual population abundance at the time. For instance, the recruitment will be most closely determined by the population at a time t_c previously, when t_c is the mean age at recruitment. Typically t_c is of the order of several years—as much as 8 years for the Pacific halibut (*Hippoglossus stenolepis* Schmidt) (Southward, 1968). One attempt to deal with this problem is that of Gulland (1961, 1968c). Here the catch per unit effort is related to the fishing effort, using not the effort in the period of observation (cf. equation (5) using catch per unit effort as a measure of biomass), but the average effort over a period equal to the mean life-span of the fish in the fishery. This procedure takes account of the fact that intensive fishing in one year will directly affect the stock for some years in the future, in fact for as long as the fish caught might have survived in significant numbers, but does not take account of subsequent effects on recruitment. It is therefore analogous to the constant recruitment models discussed in the following sections.

A similar analysis may be made when a fishery is based on several species. Ideally such a fishery requires very complex analysis, taking into account the dynamics of the population of each species, and the interaction between them. Considering how many questions remain unsolved concerning the herring in the North Sea despite lengthy research on this species, it is clearly unrealistic to expect full analysis to be made at all quickly on each species in, say, a tropical demersal fishery, where perhaps 50 species are common, and over 100 occur occasionally. On the other hand, the fishery administrator, and the fishing industry, may require some scientific advice without undue delay. A reasonable first approach may then be to consider the total biomass of all species, and the effect of fishing on it.

In the Gulf of Thailand a trawl fishery has recently developed, and has rapidly expanded to become a major fishery on the world scale (Tiews, 1965). Good measures of total stock abundance are available

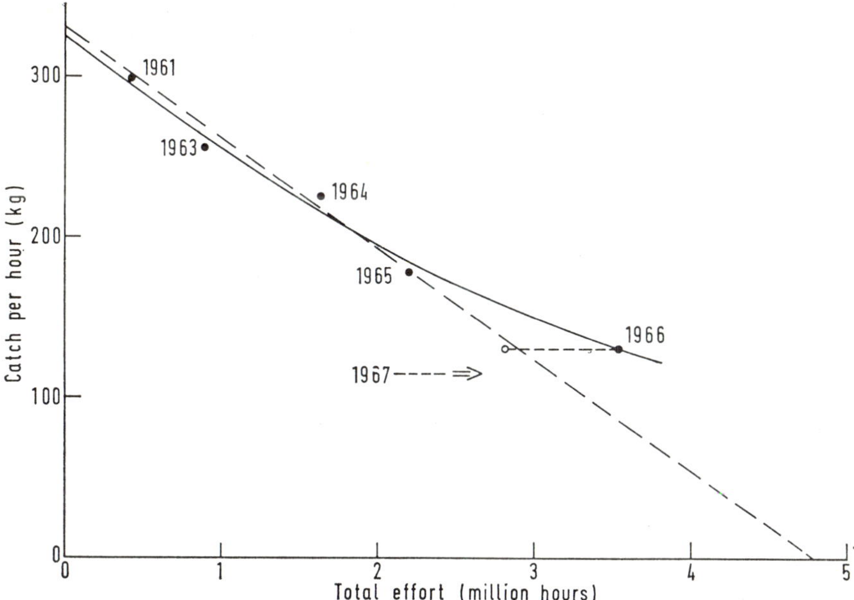

Fig. 10. The total catch per hour trawling of all species of fish in the Gulf of Thailand, related to the amount of fishing (the latter may, as suggested by a circle, have been overestimated in 1966). Full line shows fit to observed uncorrected data; broken line shows fit to data corrected for catches taken outside Gulf of Thailand in later years.

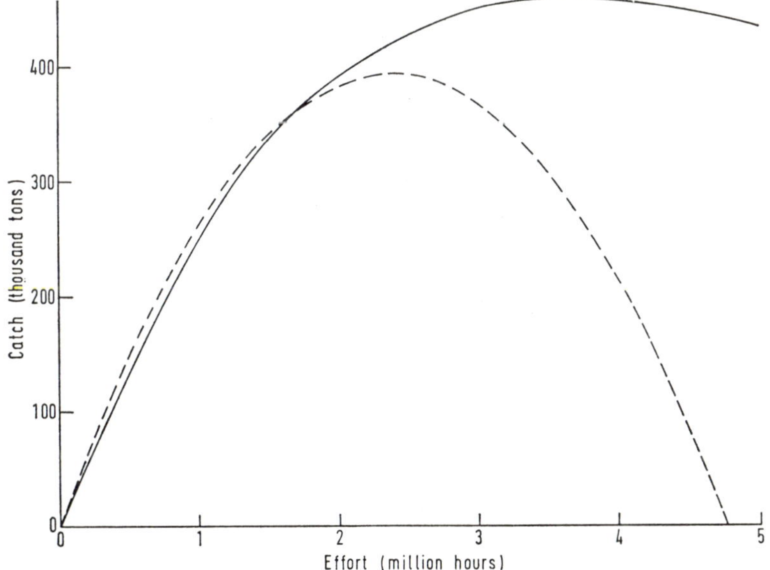

Fig. 11. The relation between the amount of trawling in the Gulf of Thailand and the total catch of all species. Full line shows fit to observed uncorrected data; broken line shows fit to data corrected for catches taken outside Gulf of Thailand in later years.

from a series of routine trawl hauls by research trawlers, which have shown a clear decline in abundance as the fishery developed (Tiews *et al.*, 1967). This decline can be clearly correlated with the amount of fishing (Fig. 10, using data from Menasveta, 1968), and from this, the relation between total catch and the amount of fishing can be deduced (Fig. 11).

Certainly this analysis gives no more than a crude picture of the events. There have been considerable changes in the species composition; while the total stock has decreased (some species very drastically), other species, e.g. various squids, have increased. However, the picture seems clear enough to be of practical value in the planning and management of the fishery; fishing has reduced the stocks taken as a whole, and it is unlikely that further increase of fishing on the grounds will give much increase in total catch.

E. Analytical Models

In these models the population is considered as the sum of its individuals, and the changes in the population are the net effects of additions by recruitment and growth, and losses by natural mortality and fishing, that is, in Russell's (1942) formulation (with slightly altered notation)

$$B_2 = B_1 + (R+G) - (M+C).$$

The earliest use of this method was by Baranov (1918) and the best known expositions are by Beverton and Holt (1956, 1957), and Ricker (1948, 1958).

In a general form, we may write

t_r = age at which fish recruit to the fishery
t_1 = maximum age in the exploitable stock
$N_t dt$ = number of fish between age t and $t+dt$ alive at a given time
W_t = average weight of fish of age t.

Then the numbers of fish in the exploitable stock, N, will be given by

$$N = \int_{t_r}^{t_1} N_t dt$$

and the total biomass, B, by

$$B = \int_{tr}^{t_1} N_t W_t dt.$$

Also the numbers and weight of fish caught per unit time by

$$C = \int_{t_r}^{t_1} F_t N_t dt \tag{8}$$

and

$$Y = \int_{t_r}^{t_1} F_t N_t W_t dt. \tag{9}$$

These equations are entirely general, and make no more assumptions about the dynamics of the population than that the population biomass is the sum of biomass of the individuals in the stock. To provide useful information, however, some assumptions must be made concerning the forms of F, N, and W as functions of age (and also in the general case as functions of time). Also, at least before the computer age, it was desirable that these functions were such that the expressions in equations (8) and (9) could be readily integrated.

F. THE YIELD PER RECRUIT

Under steady state conditions the yield in unit time from all the fish present will be equal to the yield from the group of fish recruiting in unit time during their whole life span. That is, the above expression can be considered as relating to a fixed group of recruits (a year-class). Then we can write

$$N_t = S_t R,$$

where R = number of recruits,
S_t = proportion surviving to age t

and also

$$\frac{dN_t}{dt} = -(M_t + F_t)N_t$$

so that

$$S_t = \exp(-\int_{t_r}^{t}(M_t + F_t)dt)$$

which again, is quite general, but to be useful requires some assumptions; usually these are

M_t = constant = M
F_t constant = F above some age t_c, depending on the selectivity of the gear ($t_c = t_r$ for a non-selective gear)

and

$$F_t = 0 \text{ for } t < t_c.$$

Then

$$S_t = e^{-M(t-t_r)} \text{ for } t_t < t \leqq t_c$$

and

$$S_t = e^{-M(t_c-t_r) - (F+M)(t-t_r)} \text{ for } t \geqq t_c.$$

Various assumptions have been made concerning the form of $W(t)$. This is a matter of finding a growth function which fits the observed data, and is algebraically convenient. The choice of growth functions is wide, e.g. the Gompertz (Winsor, 1932*a*; Silliman, 1966), the logistic (Winsor, 1932b), the von Bertalanffy (Bertalanffy, 1938, 1957), or exponential (Brody, 1945; Parker and Larkin, 1959), or more general forms (Richards, 1959). The most widely used in this type of analysis is the von Bertalanffy (Beverton and Holt, 1957). This gives the weight as

$$W_t = W_\infty \left(1-e^{-K(t-t_o)}\right)^3 \tag{10}$$

where W_∞ = limiting weight

K = constant, proportional to the rate at which the fish completes its growth.

Inserting these values in equation (9), gives

$$Y = \int_{t_r}^{t_1} F.Re^{-M(t_c-t_r)-(F+M)(t-t_c)} W_\infty \left(1-e^{-K(t-t_o)}\right)^3 dt$$

which on integrating becomes,

$$Y = FR\, e^{-M(t_c-t_r)} W_\infty \sum_0^3 \frac{U_n}{F+M+nK} e^{-nK(t_c-t_o)} \times \left(1-e^{-(F+M+nK)(t_1-t_c)}\right) \tag{11}$$

where U_0, U_1, U_2, U_3 are constants for the summation, and are equal to 1, -3, 3, -1 respectively.

Ricker (1948, 1958) used an exponential growth curve, which may be written

$$W_t = W_c e^{g(t-t_c)} \tag{12}$$

where W_c = weight at time t_c.

The equation for the yield then becomes

$$Y = \int_{t_c}^{t_1} FR\, e^{-M(t_c-t_r)-(F+M)(t-t_c)}.\, W_c e^{g(t-t_c)}\, dt$$

or

$$Y = FR\, e^{-M(t_c-t_r)}.\, \frac{W_c}{F+M-g} \left(1-e^{-(F+M-g)(t_1-t_c)}\right). \tag{13}$$

This growth curve does not usually fit observed data over such a wide range of ages as the von Bertalanffy, but the span of ages in the fishery may be divided into a number of intervals, in each of which the fit is good.

Such a division, into possibly a large number of intervals, and the associated computation, can be quite easily achieved with modern computer facilities (Doi, 1962; Silliman, 1966; Paulik and Bayliff, 1967). Computers can in fact be used to calculate the yield for any form of growth. Either a specific algebraic formula can be used, or the observed data on weight-at-age, which latter can include, if desired, seasonal variations in growth rate, or condition factor.

Although these equations contain several parameters, only two can be immediately controlled by the fishermen—the size at first capture and the fishing mortality, which will be proportional to the amount of fishing if this is correctly measured. Then the relevance to the fishery may be given as two sets of curves, one of catch as a function of fishing mortality for fixed sizes at first capture (cf. Fig. 2) and the other of catch as a function of size at first capture for fixed values of fishing mortality. Alternatively both relations can be shown in a single isopleth diagram (Beverton, 1953; Dickie and McCracken, 1955; Beverton and Holt, 1957).

Because the recruitment R is rarely known in absolute terms the yield as such is rarely calculated, but rather the yield per recruit (Y/R). Provided the recruitment is constant this will of course be proportional to the yield, and for example the mesh size of trawls (which control the size at first capture), which maximizes the yield per recruit, could also maximize the total yield.

Another reason for expressing the advice to the fishing industry in terms of yield per recruit is that in several fisheries the recruitment fluctuates greatly, and apparently independently of the adult stock and of the amount of fishing. In such a fishery advice that, say, increasing the trawl mesh size from 90 mm to 110 mm would increase the yield of haddock by 6% would be quite misleading in the short run, if immediately after the change a number of weak year-classes recruited to the fishery. However, advice that the yield per recruit would increase by 6% would be correct, and would give a reliable guide to the improvement achieved in the catches, compared with what would have happened if the mesh size had not been increased (Beverton and Hodder, 1962).

Computers can be used to generalize the expression above. Thus it is not necessary to assume the above simple form for F_t (equal to 0 or F), but the fishing mortality can be varied in accordance with the season or the age (and size) of the fish. A complex model (Royce *et al.*, 1963) of the salmon fishery on the American Pacific coast takes account of the movement of salmon along the coast, and of the variations of the fishing effort on them at each point in accordance with the success of the fishing, and the regulations imposed. Conceptually the important distinction, however complex the computer programme, is between models in which the natural parameters of growth, natural mortality and recruitment are independent of adult stock (as in the equations above) and those in which one or other parameter is a function of stock abundance, as is discussed in the following sections.

IV. More Complex Models

In the previous section, it was assumed that the only effect of fishing was the direct effect of an element added to total mortality, and that the other parameters (growth, mortality, or recruitment) were not affected. Fishing cannot affect these parameters directly, but it is almost certain that as fishing increases, and reduces the abundance of the exploited population, there will be indirect effects on the parameters.

Formally such changes can be taken into account by expressing each parameter as a function of the population

i.e.
$$M = M(P)$$
$$R = R(P) \text{ etc.}$$

These expressions can then be combined with the earlier equations and by more or less complex mathematics, or by successive iteration (which good computing facilities are making increasingly easy) the population abundance, yield, etc., can be determined for any pattern of fishing.

However, though this description of the theoretical solution to the problem is easy enough, it is of little practical value without some guidance in the form of the various functions $M(P)$, $R(P)$, etc., and the ability to fit these functions to the conditions of the fish stock of interest. This has proved very difficult, and perhaps the most useful practical progress made so far is the examination of the implications of various departures from constancy of growth, mortality and recruitment, and the effects on the conclusions from the constant-parameter model of the more likely departures from constancy. The more extensive of such studies have been those of Beverton and Holt (1957).

A. density-dependent mortality

Natural mortality is the most difficult to measure, except in the rather unusual case of detailed examination of an unexploited population—in which the total mortality (which may be estimated from the age-composition of research samples) is equal to the natural mortality. In exploited populations natural mortality is usually estimated from subtracting the more easily measured fishing mortality from the total mortality (as estimated from age-composition data, or from tagging experiments). Commonly the two sources of mortality are separated by plotting the regression of total mortality on fishing effort (Fig. 12). The assumption made in using this method is that

$$Z = F+M = M+qf$$

so that if M is constant, the relation between f and Z is linear with intercept on the Z axis equal to M. Clearly this method cannot determine whether there are any changes in M, and only a single estimate of M is obtained, which, if M varies, will probably be fairly close to the value of M in an unexploited population.

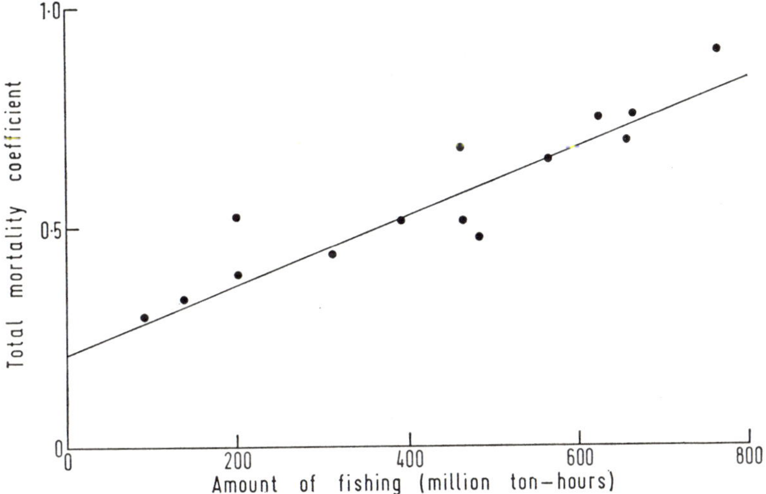

Fig. 12. The relation between the amount of fishing on the Arcto-Norwegian stock of cod, and estimated total mortality coefficient among 6-year-old fish (data from ICES Arctic working group).

There are few if any clear examples of changes in natural mortality—except for the occasional outbursts of extremely high mortality owing to "red tide", disease (e.g. in Gulf of St Lawrence herring, Sindermann, 1958) and unusual cold (Lumby and Atkinson, 1929; Storey and Gudger, 1936; Templeman, 1965)—and of these none seems to be clearly related to changes in abundance. A possible example of a change in natural mortality over a period is that of the Californian sardine (*Sardinops caerulea*) where it appeared to increase from 0·4 in the 1930's to 0·8 in the 1950's (Murphy, 1966), and over this period the stock abundance decreased greatly.

By making the assumption that natural mortality was linearly related to the population numbers, Beverton and Holt (1957) showed that the shape of yield-effort curves based on density-dependent natural mortality differed little from those based on constant mortality except at very low (and for practical purposes usually less interesting) levels of fishing effort. Such differences as there are tend to make the curve flatter than when mortality is constant. It may be that the variation of natural mortality with density is not the simple function assumed by Beverton and Holt, e.g. mortality caused by some disease may be low and fairly constant until a critical population level is reached, and may then become large. However, it does seem that there is some justification, other than the great difficulty and inconvenience of doing otherwise, for ignoring density-dependent variations in natural mortality of the exploited stock (although, as noted below, most of the models concerned with the stock-recruitment relationship imply considerable density-dependent mortality among the young fish or eggs).

B. DENSITY-DEPENDENT GROWTH

Changes in growth are easier to observe. A well-behaved fish, such as a herring or salmon (*Oncorhynchus* spp.), bears on its scale evidence not only of its age but of its size (length) at the end of each year of its life. Thus even one fish can provide some evidence concerning year-to-year changes in growth. The observed data of length-at-age may be difficult to interpret directly since the size of a 6-year-old fish in 1968 will be determined by the sum of the growth in each of the years 1962 to 1968. Given data either of the detailed growth history of individual fish, or of the mean length-at-age for the population each year, the information can be expressed as the growth increment between say 1966 and 1967. This can then be fitted by a suitable growth curve, and year-to-year changes in the parameters of this equation can be studied (Fig. 13). The von Bertalanffy is usually a convenient growth curve to use, especially if, following Beverton and Holt (1957), it is assumed that density-dependent growth changes will not affect the value of K, which remains constant. Then there will be only a single parameter, L_∞, which varies, and this can then be directly correlated with any suitable measure of stock abundance. Although Beverton and Holt made the assumption of constant K in their analysis of the growth of North Sea haddock, a re-examination of the data, from Raitt (1939), which they used (Gulland, 1962) showed that the clearly observable changes in the growth increment in the second year of life could have been due to changes in K, or L_∞, or both, and there was no evidence that K did in fact remain constant.

Clearly defined changes in growth have been observed in several fish stocks, especially in temperate and sub-arctic waters where growth and age determination is easy, but the relation of these changes to stock abundance has not always been clear. A change that appears to be a

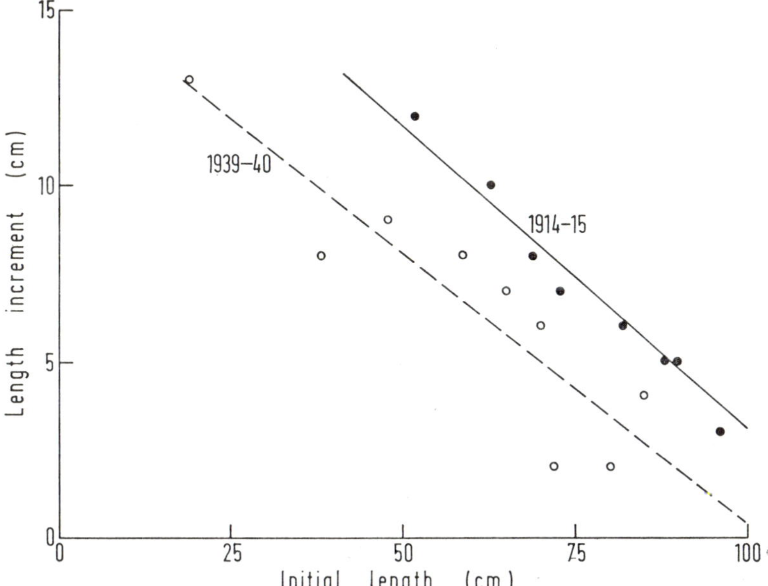

Fig. 13. Changes in the annual growth of Pacific halibut, showing the increments between 1914 and 1915, and 1939 and 1940, as functions of the length in the first year, and straight lines drawn by eye (data from Southward, 1968, Table 3).

change in growth related to density has occurred in Antarctic whales. Figure 14 shows the length composition of female fin whales caught by pelagic operations in the Antarctic during the seasons 1945/6 and 1963/4 (data from International Whaling Statistics, XVIII and LII). One difference is the much bigger proportion of small whales in the recent catches; this is partly a real change—fewer whales are surviving to be full-grown—and partly due to the gunners in the catcher vessels avoiding the small whales in the 1945/6 season when whales were more abundant. (The small mode at 57 feet, the size limit, is due to the well-known "elastic" measuring tape used when the whale is just below the minimum size.) The other difference is a difference in the mode among the bigger animals—70 feet in 1945/6 and 73 feet in 1963/4. The shape of the length distribution at this size probably reflects more than anything else the variation in the ultimate size (the L_∞ of the von Bertalanffy curve) between different animals. Though in the absence of reliable age-determinations from the earlier period, the evidence is not conclusive, it seems likely that there has been a real change in the growth of fin whales, which is correlated with the great decline in abundance of fin whales (and blue whales) during the same period (Chapman, 1964).

This scarcity of clear evidence relating growth to population density is in part due to the different segments of a fish population behaving

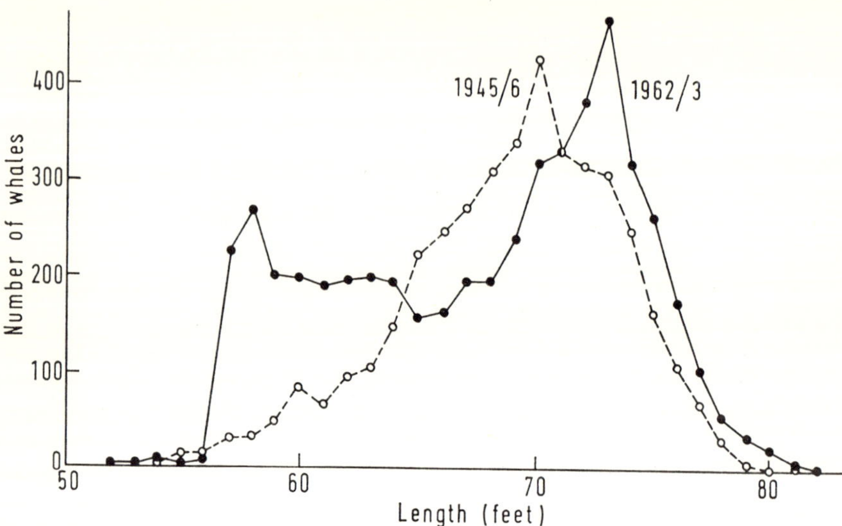

FIG. 14. Length composition of female fin whales caught in the Antarctic in the 1945/6 season (open circles), and 1962/3 season (closed circles), showing mode at a longer length in the latter season, possibly due to better growth.

independently. Thus under heavy exploitation the exploited part may decrease, while the abundance of the pre-recruits remains unchanged. Some of the clearest evidence of density-dependent growth has been found in the early years of life of fish with great variations in year-class strength, e.g. in the North Sea haddock referred to above. Another example is the cod in the Barents Sea. Ponomarenko (1967) has shown that in the most recent years, when the exploited stocks have been at a low level, the sizes of most of the age-groups in the catch have been above average. Detailed examination of the data shows that the biggest difference is at about 4-5 years old, when the fish are recruiting to the fishery, and that thereafter the growth increments are about normal (thus maintaining the differences in size at 5 years old). The real change in growth is, therefore, limited to the younger fish, perhaps correlated with the weak year-classes which have recently occurred, but the intense fishing, which has reduced the abundance of larger fish, has had no effect on growth.

Growth changes are more readily susceptible to analytical studies than mortality changes, since they can be linked to the consumption of food by the animal, and the abundance (and population dynamics if desired) of the food species; the concept of Ivlev (1961) on the feeding of fish, and the relation of the quantity consumed to hunger, and the availability of food, are most important to such studies. Equally important are studies of the effects on the food population of changes in the abundance and food consumption of the predator population. However, like other research problems, these studies have rarely been made in the sea. The clearest example of the reduction of a prey population was the collapse of the lake trout (*Salvelinus namaycush*) in several of the Great Lakes following the arrival of the sea lamprey (*Petromyzon marinus* (Applegate, 1950). This experience is perhaps not very helpful to the general problem because it demonstrates the effect of a new predator rather than of the increased abundance of a long-established predator.

Density-dependent growth is well known in stocking ponds, where very high densities can occur (Walter, 1934; Le Cren, 1965; Backiel and Le Cren, 1967). In some ponds increasing density of stocking of carp (*Cyprinus carpio*) changed the composition of the plankton and benthos in the pond, actually tending to increase the plankton population (Hrbacek *et al.*, 1961; Grygierek, 1962, 1965). The stocking rates were very high (up to 30 000 fish/ha), so that the effects were both more complex (including quicker re-cycling of nutrients, and higher primary production at the high fish densities), and more readily detected than among natural fish populations in the sea or in fresh water. In such conditions of high density other subtle factors—the "conditioning" of the water (Allee *et al.*, 1940) or hierarchies (Brown, 1946)—can affect the growth, while they are unlikely to be important in natural waters.

While the exact relation between growth and population density may remain uncertain, it is possible, as for density-dependent mortality, to examine the effect of reasonable hypotheses on the relation between catch and fishing effort. Thus Beverton and Holt (1957) examined some half-dozen different hypotheses. All tend to have the same effect—at high densities growth is slowed down, and at low densities growth is quicker, thus tending to flatten out the curve. This is similar to the effect of introducing a density-dependent mortality. In both cases advice to the fishing industry or administrator will be unaltered—action that will increase the total catch will still increase it, but perhaps not so much as expected from the constant parameter model, and action that will decrease the catch will still decrease it.

C. DENSITY-DEPENDENT RECRUITMENT

The more likely density effects on recruitment—that recruitment will be low at low adult densities, and high at high densities—are more critical to any conclusions and advice than density-dependent growth or mortality, since recruitment changes tend to exaggerate any differences in catch from different actions, or to alter the nature of the conclusions. For instance, the model based on constant recruitment may predict for a moderately heavily fished stock some small increase in total catch from increased fishing, whereas on reasonable assumptions about the stock density/recruitment relation, increased fishing may result in a catastrophic fall in total catch. The relation between adult stock and subsequent recruitment has therefore received considerable attention from fishery scientists (Ricker, 1954, 1958; Beverton and Holt, 1957; Cushing 1968).

The simplest assumption, and the one made in the constant-parameter model above—that the average recruitment is constant and independent of adult stock—itself includes a very definite implicit assumption concerning the effects of adult density on the mortality among the early, pre-recruit stages. The number of eggs laid, E, is closely proportional to the abundance of the adult stock, S, or more strictly to the biomass of adult females. If the number of recruits, R, is constant, the probability of the young fish from an individual egg surviving to the age of recruitment is R/E, i.e. inversely proportional to the number of eggs produced and, therefore, also to the abundance of the adults. This is shown clearly in Figs 15 and 16, for the North Sea plaice stock, in which the total recruitment appears to have remained remarkably constant (Beverton, 1962).

The other simple population model, the Schaefer model, also makes clear (if implicit) assumptions concerning the stock/recruitment relation. If it is assumed that the other parameters remain constant, it is a matter of simple arithmetic to determine what stock/recruitment relation is required to make the results from a constant-parameter yield-per-recruit calculation agree with the estimates of total yield calculated by applying the Schaefer model. Schaefer (1967a) has done this for the yellowfin tuna of the eastern tropical Pacific. The resulting curves (his Fig. 6) are dome-shaped, and very similar to the curves proposed by Ricker (1954, 1958) although closer analysis showed some differences.

These calculations do not establish the true relation between stock and recruitment, but do indicate how, implicitly or explicitly, any analysis of the dynamics of a fish stock must make some assumption concerning the stock/recruitment relation. Different assumptions can make very large differences to the scientific conclusions and to the advice given by the scientists to the fishermen. Thus in the world's biggest single species fishery, for anchovy (*Engraulis ringens*) off Peru, it is quite clear that fishing is taking a large part of the stock. Doubling the amount of fishing would increase the yield per recruit by perhaps 20%; if the average recruitment is not affected, then the doubled effort would produce nearly 2 million tons more fish. However, on the Schaefer model doubled fishing, if maintained, would cause the collapse

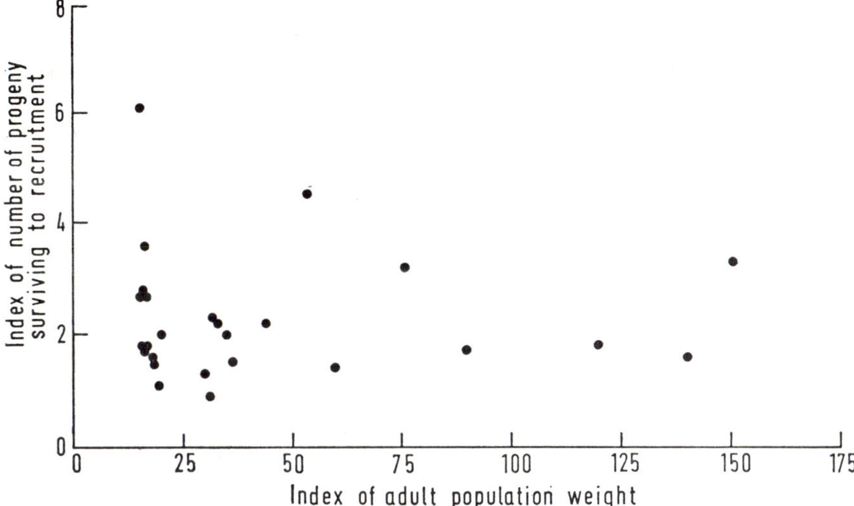

Fig. 15. Year-class strength of North Sea plaice plotted against the abundance of the parent stock, showing absence of correlation (from Beverton, 1962).

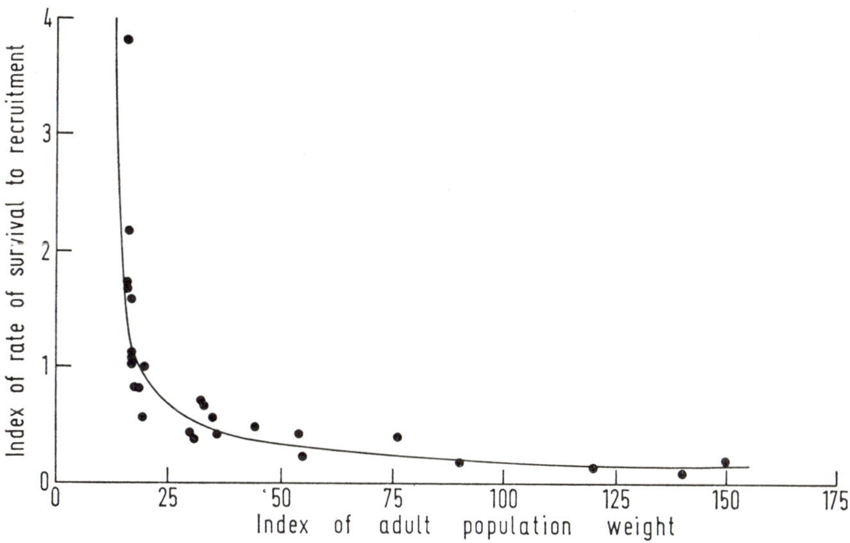

Fig. 16. Index of survival of North Sea plaice from eggs to recruitment, showing reduced survival at larger parent stocks (from Beverton, 1962).

of the stock, and the loss of most of the catch (8–10 million tons per year) (Schaefer, 1967b). Unfortunately, only seven pairs of values of adult stock and subsequent recruitment are as yet available, and the plot of recruitment against adult stock is inconclusive; it could, not unreasonably, be interpreted as showing that further moderate decreases in adult stock would increase the recruitment (Gulland, 1968d). This implies that a doubled effort would increase the catch by even more than the 2 million tons estimated on the constant recruit model.

Although the uncertainty over the Peruvian anchovy—a range of perhaps 10 million tons—is the most dramatic and the clearest example among present fisheries, similar uncertainties are occurring or have occurred in many other heavily exploited stocks. While uncertainties still exist concerning the collapse of the fishery for the California sardine, the catches of which fell from a peak of 800 000 tons in 1936/7 (it was then among the world's biggest fisheries) to zero now, it seems fairly certain that it was a recruitment failure, caused initially by a fishery-induced reduction in adult stock, but added to by competition with anchovy (*Engraulis mordax*) (Murphy, 1966). At the time when the stock was declining, but still large, there was evidence that the decline might be due solely to environmental effects (Clark and Marr, 1955; Marr, 1960). Largely because of these uncertainties, no effective action was taken to maintain the adult stock (or reduce the anchovies) at the time when action might have been productive, and the sardine fishery is now non-existent.

The importance of understanding the stock/recruitment relation to the practical problem of fishery management needs no further emphasis. At the same time it is of major interest purely as a scientific problem.

Following the methods used in the analysis of the exploited stock, R may be expressed as a function of the instantaneous mortality coefficient M acting during the pre-recruit period. This will not be constant throughout the period, but can be written as a function of age,

$$\frac{dN}{dt} = -M_t N$$

where N = number of young surviving to age t

and $$R = E \exp\left(-\int_0^{t_c} M_t dt\right) = E\, e^{-y}$$

where E = number of eggs produced,

$$y = \int_0^{t_c} M_t dt.$$

If M_t were independent of the size of the spawning stock, then y would be a constant, i.e. the recruitment will be proportional to the egg production and to the adult stock. Certainly M_t will be in some way related to the adult stock, and by making various assumptions about this relation, any desired stock/recruit curve can be produced, given sufficient mathematical ingenuity.

Thus, following Ricker (1954), M_t may contain, in addition to an element independent of stock abundance, an additional element proportional to the adult stock (e.g. resulting from cannibalism), i.e.

$$M_t = a_t + b_t E$$

and
$$y = -\int_0^{t_c} M_t dt = -\int (a_t + b_t E) dt = -c - dE$$

where c, d are constants.

Therefore $R = Ee^{-c-dE}$.

This produces an asymmetrical dome-shaped curve, with a definite maximum.

Alternatively, following Beverton and Holt (1957), M_t may have an additional term proportional to the number of young fish alive at the instant being considered

$$M_t = a_t + b_t N$$

and
$$\frac{dN}{dt} = -(a_t + b_t N)N.$$

This equation is not easy to solve explicitly, but by considering a number of intervals, in each of which a_t and b_t might be taken as constant, Beverton and Holt showed that the resulting egg-recruit relation was of the form

$$R = \frac{E}{aE + b}$$

which gives a curve that rises to an asymptote ($R = {}^1/a$) and has no maximum.

Neither of these simple mathematical models is likely to be very close to reality in more than a few situations. Ricker's formulations can give both a good fit and a reasonable explanation to the data obtained by Silliman and Gutsell (1958) for guppies (*Lebistes reticulatus*) in laboratory populations, where cannibalism is known to occur. Cannibalism is however not common in many commercial species; young hake (*Merluccius merluccius*) may be a substantial element in the food of large hake (Hickling, 1934) and large cod may eat some small cod, but in other species, e.g. plaice, young and adult are to a considerable extent geographically separated (Wimpenny 1953).

Equally, it is unlikely that mortality is related solely to the abundance of young at the same instant of time, and the events of the preceding period are also important. The most persuasive explanation of the stock/recruitment phenomenon is that the most important factor is food supply to the young. Certainly the food supply can affect the growth and condition of larval fish (Shelbourne, 1957) and if unable to increase mortality directly, can do so by causing greater vulnerability to predators.

Increased numbers of young fish mean increased grazing on the food supply, a reduced food supply a little later, and then increased mortality. If food supply is important, the contemporary population density of young fish would have rather less influence on the mortality than the population in the immediately preceding periods.

Under these conditions, it is clearly possible to devise a model which would predict zero recruitment above a certain initial adult stock—an initial density great enough to graze the food supply down to a level at which the young fish would all starve. Clearly such a model is less realistic than the two previous ones. It would require, for instance, that the period needed for the larvae to starve was substantially less than the period for the food population to recover, but it does illustrate the range of possibilities open. Such considerations of the food supply also suggest that a stock/recruitment curve with a maximum is quite feasible even when cannibalism does not occur.

D. YEAR-CLASS FLUCTUATIONS

The application of the Ricker or Beverton and Holt or other model to a particular fishery is made difficult by the commonly very great fluctuations in the number of recruits, which are clearly not related in any simple way to adult stock. In temperate and sub-arctic waters the fluctuations in year-class strength can be very great. The single 1904 year-class of Norwegian herring was about three times as large as that of the next largest year-class between 1899 and 1920, and perhaps twenty times the average of the rest. More recently the year-classes of haddock on 1962 and 1963 on Georges Bank, and 1962 in the North Sea were all several times the average, and at least 100 times the strength of some recent weak year-classes (Jones, 1966; Graham, 1967). The haddock catch on Georges Bank rose to 155 000 tons in 1965, compared with catch under best management from year-classes of average strength of 45 000 tons; in the North Sea the haddock catch rose from under 60 000 tons in 1963 to 260 000 tons in 1966. The size of the spawning stocks from which these outstanding year-classes originated was by no means abnormal, and very similar to the size of the parent stock from which very weak year-classes originated at a separation of only a few years. Subsequent year-classes on Georges Bank have, up to 1968, all been very weak, but those in the North Sea have been good, one even better than 1962.

These year-class fluctuations produce a great scatter on any plot of R against E (or R/E against E, or similar derivations), so that the fitting of a curve to observed data is difficult. This difficulty is increased by the fact that the number of pairs of points available from n years' observations is only $n-x$, where x is the earliest age of fish at which it is possible to estimate year-class strength.

This shortage of data (no more than 1 point per year) has inhibited studies of the causes of these year-class fluctuations. Numerous attempts have been made to correlate year-class strength with various environmental factors, and different authors reaching often quite different conclusions (e.g. Corlett, 1965; Gulland, 1965). The passage of time is the surest, if very slow, way of resolving such differences, and has indeed shown many such correlations to be either unfounded, or founded on a transitory situation, and not applicable to data covering a long period. Analysis by normal statistical techniques is often not helpful, for the range of possible environmental factors to consider (temperature, salinity, etc., for different areas at different seasons) is very large so that some element of selection is inevitable. Lacking any objective criterion of selection, this is inevitably done to ensure a good correlation; a reasonably good correlation is therefore almost inevitable, and the normal tests of statistical significance are not applicable (Gulland, 1953).

In the simplest conditions the fluctuations in year-classes might be considered as a simple additive or multiplicative random term in the relation between stock and recruits, i.e. in the Ricker model,

for constant environment $R = Ee^{-c-dE}$

or $\log R = \log E - c - dE$

then for varying conditions we might write

$$R = Ee^{-c-dE} + x$$

or $\log R = \log E - c - dE + x'$

where x, x' are simple random variables, mean 0.

Such variations would not affect the nature of the stock/recruit curve to any important extent—for instance the value of E that gives the maximum value of R would be unchanged. It is however quite likely that the effects of the environment will be more complex, for instance affecting the value of d in the equations above, so that it is necessary to write

$$\log R = \log E - (c+x) - (d+y)\, E$$

where x, y are random variables.

This supposition is reasonable, since the value of d depends on the competition among the young fish; certainly if this value is determined by food, it would be expected to vary inversely with the supply available. Variations in d (and analogous variations for other models of stock/recruitment) would make profound changes in the shape of the curve, e.g. in the position of the maximum. These effects make it still more difficult to establish the true relation between stock and recruitment, or between recruitment and any particular environmental factor, and in particular render inapplicable such useful techniques as multiple correlation.

Just as the simple study of direct effect of fishing on the exploited stock by correlating fishing effort and catches has often proved inadequate, and has to be replaced or strengthened by detailed analysis of mortality rates, etc., so the more conclusive studies of the factors influencing recruitment will come from direct observations on the mortality rates (and also growth rates, etc.) among the young fish. In such a study the more relevant relation is between R/S (the index of survival) and the total stock, S rather than between R and S, though in making any interpretation of a plot of R/S against S one must guard against the danger of a spurious relation being introduced purely by the arithmetic of the procedure. The study of the population dynamics of the young fish is much more difficult than the study of the exploited ages. Many of the advantages that have so greatly assisted the study of the exploited stock—extensive data from the commercial fisheries, often reliable age data, etc.—are absent.

Some studies have however been carried out (e.g. Sette, 1943; Ahlstrom, 1954, 1965), based on extensive plankton surveys by research vessels; of these the CALCOFI surveys (instituted largely as a result of the collapse of the sardine population noted above) have been outstanding. These surveys have produced estimates at fairly frequent intervals of the number of egg larvae of different sizes and at different stages of development. Equating these data with age enables curves analogous to the age-distribution of the exploited stock to be drawn. These curves suggest that in the early months of life the mortality is very high, but are without any clear indication of abrupt changes in this mortality, or of periods when mortality is particularly high.

The big question hanging over this work is, however, not whether the mortality can be estimated, but whether it can be estimated accurately enough for real changes in mortality to be detected and related to population density or environmental factors. So far as the environmental effect at least is concerned, the precision required will depend on whether the differences in mortality which affect year-class strength are large differences, concentrated in one short period (the critical period of Marr, 1956) or small differences lasting for much of the larval or post-larval life.

Apart from the large amount of sea work and subsequent analysis that such a programme involves, laboratory studies concerning such things as food requirements, growth rates, etc., of the larval fish are also likely to be required. The correlation observed in the field between apparent mortality and density of food is rather weak. Clearly this would become much more convincing if the mortality in the field became appreciable at a food density below which increased mortality or appreciably reduced growth also occurred in the controlled conditions of a laboratory.

Much of this, however, is speculation. The present situation is that a better knowledge of the relation between adult stock and average recruitment is desirable but not available for most marine stocks, nor is there any clear formula for obtaining this knowledge.

V. Factors affecting Total Abundance

A feature of the world's fisheries is the widely different production from different stocks—10 million tons of Peruvian anchovy, 1 million tons of North Sea herring, 100 thousand tons of shrimp from the Gulf of Mexico and 10 thousand tons of Atlantic salmon, and so on, down to species and stocks which can support only extremely small and local fisheries. The previous section has described models of changes in production and stock abundance, especially those changes caused by man's activities, but these changes (except for the occasional disaster, e.g. the California sardine) are small compared with the orders of magnitude of the differences noted above. The models, in their simpler forms at least, do not explain why the changes occur at the levels observed, rather than there being for example 10 million tons of Atlantic salmon.

One important factor is the position of the species in the food chain. It is a commonplace that the plants in the sea (except for seaweeds around the margins) are too small to harvest, and so also are most of the herbivores. Most of the commercial fish are first- or second-stage carnivores, and the order of volume of production in the list above corresponds quite closely to the order of closeness of the fish to the primary production. The anchovy is to a large extent, but by no means exclusively, a phytoplankton eater (de Ciechomski, 1967); herring eat *Calanus* and other small zooplankton animals, which eat phytoplankton, while, at least off West Greenland salmon eat sandeels and capelin, which eat zooplankton.

The position in the food chain is not the only factor affecting total abundance; more than one species may occupy a somewhat similar position; the efficiency of transfer across one link in the food chain is not likely to be constant and the original primary production is also variable from place to place.

The variation in primary production is not so great in the sea as it is on land (except for the permanently ice-bound arctic waters, there are no areas in the ocean where production is virtually zero, as is the case in the extreme desert areas) but the variation is still large. Shortage of water is no problem, and temperature is limiting only under the polar ice, but the supply of nutrients is most important and light has a controlling influence in the timing of production in boreal and arctic waters. Koblents-Mishke (1965) has summarized the information on the distribution of primary production in the Pacific (see Fig. 17). Many of the areas of high production coincide with important fishing grounds, e.g. the high production off Peru. This coincidence is no recent phenomenon. Cushing (1968) has pointed out how closely the distribution of catches of sperm whales taken by New England whalers as plotted by Townsend (1935) resembles the distribution of primary production throughout the world.

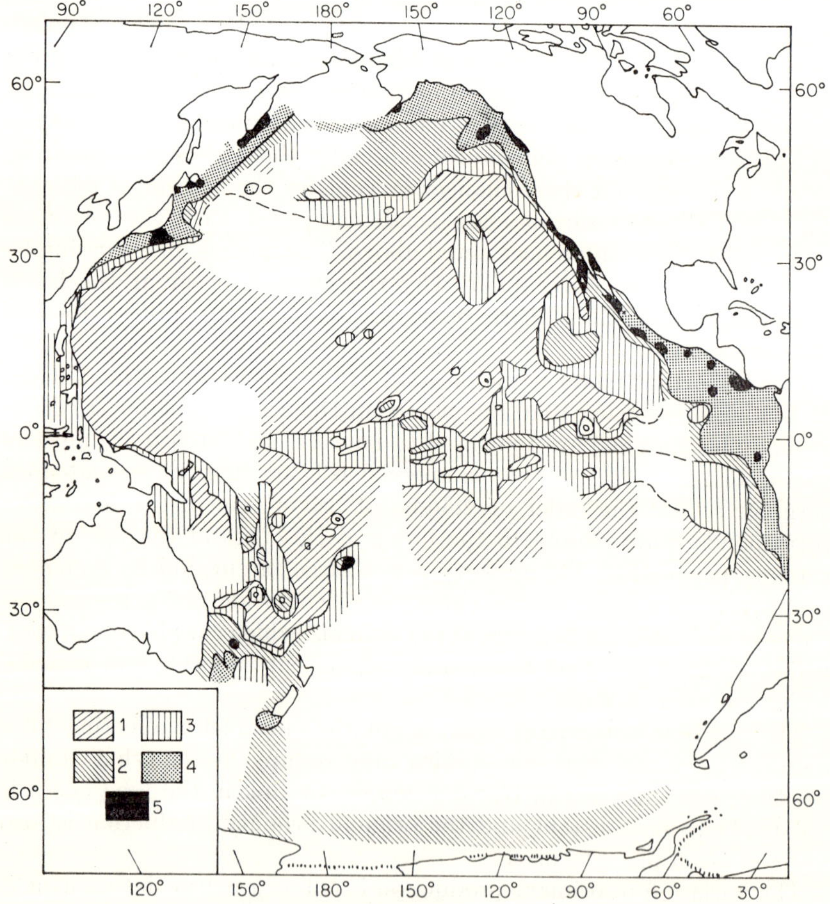

Fig. 17. Distribution of primary production in the Pacific (from Koblents-Mishke, 1965).

A high annual production of fish is less interesting to the fishermen than high standing stock, and the relation between standing stock and production varies regionally. In the tropics, the rate of turnover is high, so that the standing stock may be equal to the production for only one year or less; in unfished stocks in temperate or boreal waters fish may live for twenty years or more and the standing crop may be equal to the accumulated production over ten years.

Another advantage to fishermen in the temperate and sub-arctic waters is that the number of species is less; several great fisheries are based on single species, such as the cod or the herring, whereas a single trawl in the tropics may yield several dozen species. A small number of species clearly simplifies marketing, even though the widespread view of the poorer flavour of tropical fish probably owes at least as much to the greater difficulties of adequate preservation as to any poorer flavour at the time the fish leave the water.

The earliest long-range fisheries, e.g. the cod fisheries on the Grand Banks of Newfoundland which are several centuries old, have therefore been in temperate or sub-arctic waters. As these stocks have been fished down, fishing has spread all over the world—Japanese long-line vessels now fish for tuna in all the oceans of the world; Russian trawls took nearly half a million tons of hake (*M. hubbsii*) in the south-west Atlantic in 1967—and as the standing crops become reduced, the distribution of fishing becomes, on a world scale, increasingly similar to the distribution of primary production, except that fishing is still generally confined to the waters adjoining and over the continental shelves.

As well as the magnitude of the primary production, the proportion of this production which passes through the food chain and appears as fish influences the magnitude of the fish production. A common assumption is that the production of one trophic level is around 10% of the production of the previous trophic level. This is the ecological efficiency, sometimes defined as production at one level expressed as a percentage of the total consumption by the organisms at that level. There is some evidence that this efficiency does not vary much (Slobodkin, 1962), and it has often, and most conveniently, been taken as being 10%.

The discussions at the recent symposium on Food Chains in the Sea (Steele, 1970) suggested that matters are not quite so simple. The ecological efficiency is the product of three quite independent elements —the proportion of the food consumed that is used for growth, the efficiency with which this food is converted into growth, and the efficiency with which the population as a whole grazes the food population. There is no reason to suppose that this product should remain even approximately constant and some events will act in opposite directions on the different elements.

Paloheimo and Dickie (1966) have shown that the efficiency with which the food consumed by a fish is used for growth decreases with size, so that the first two elements will be greater when a fish population is composed mainly of small individuals. Such a population is produced by heavy fishing, which also reduces the total stock, and therefore decreases the ability with which the stock can graze on the food population available to it. Thus heavy fishing can reduce the overall efficiency; for example the efficiency with which the plaice and other demersal fish stocks in the North Sea used the available food was low during the period of heavy fishing in the 1930's (Gulland, 1970). The efficiency, however, would be increased if the average size of fish were reduced, but the stock is maintained by an increased recruitment (Paloheimo and Dickie, 1970). This increase occurs only if the stock recruitment curve is of an appropriate shape. As already noted, information on the stock/recruitment relation is commonly very imprecise, but there is no suggestion that it is commonly of the form—increased recruitment from reduced adult stock—which would tend to maintain the food consumption at a constant level.

There is indeed evidence to the contrary for some stocks. In the North Sea haddock recruitment fluctuates widely; the growth of the younger (1–2-year-old fish) is negatively correlated with the strength of the year-classes (Beverton and Holt, 1957, using data of Raitt, 1939). Combining data of growth of the individual and year-class strength an index of the total annual weight increment of the year-class can be computed (Gulland, 1970). Although the data do not cover the full range, it appears that this increment is small for small year-classes (as it must be), and also for extremely large year-classes (because the individuals grow very little) and that the maximum increment is at a fairly high year-class abundance, an abundance that is noticeably higher than the average year-class strength (Fig. 18). In principle the total food consumption of a year-class could be computed in the same manner though much of the basic data are lacking. However, the proportion of the total food consumption that is used for maintenance (rather than growth) will increase as the growth decreases. The abundance resulting in maximum food consumption will then be greater than that giving the maximum growth increment, and therefore *a fortiori*, greater than the average year-class abundance.

These calculations show that, under average year-class conditions, the consumption of food by a brood of young haddock is less than it might be, and that the greatest efficiency of utilization of the food resources is only achieved when quite large year-classes occur. Similar

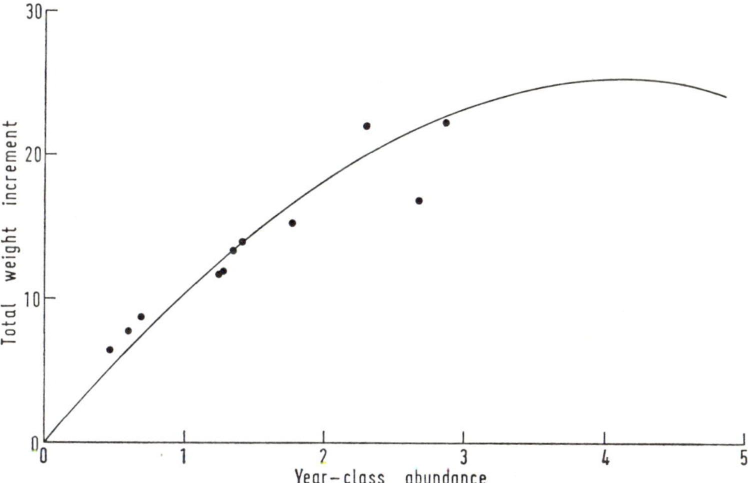

Fig. 18. Total weight increment of year-classes of North Sea haddock during their second year of life, as a function of year-class abundance.

conclusions, that the efficiency is variable, and normally less than maximum, can be reached concerning the effects of fishing. Reviewing data from the North Sea, principally for plaice, Gulland (1970) showed that the increases of growth that occur in periods of heavy fishing were not sufficient to balance the reduced numbers, so that fishing reduced all of the following—total numbers, total increment of growth (in weight) among the exploited stock, and total food consumed. Thus increased fishing decreases the efficiency with which the important commercial stocks of demersal fish (plaice, cod, haddock, etc.) utilize their food supplies.

These analyses suffer from being in essence descriptive and qualitative, since there are no field studies of how much the individual haddock or plaice eats under normal conditions in the sea, or of the efficiency with which this food is converted into growth. It might appear from the above analyses that under typical conditions of moderately heavy fishing and not more than average year-class strength this efficiency would be low, e.g. an ecological efficiency rather less than the 10% assumed as a general figure.

In fact, the ecological efficiency in the North Sea is higher than 10%. Steele (1965) and later Gulland (1970), using slightly different methods of assessing the fish production, compared estimates of primary production and of fish production. Primary production is around 100 $gC/m^2/year$. Allowing for the proportion of the fish production that dies naturally and is not recorded in man's catches, statistics of fish landings can be converted to the same units as primary production. This gives a value of fish production of $1gC/m^2/year$, in round figures, of which about two-thirds are zooplankton feeders (mainly pelagic fish, especially herring), and one-third are predators and benthos feeders (cod, plaice, etc.)

None of the major fish species in the North Sea are phytoplankton feeders, and some, e.g. the larger cod, are important predators. Thus the fish are at least two stages removed from the primary production; the fact that the fish production is around 1% of the primary production suggests, as Steele points out, that the fish stocks cannot, as a whole, be removed by more than two stages from the primary production, and that the ecological efficiency cannot be less than an average of 10% at each of the two stages.

The results also imply that the ultimate controls of the fish abundance are the primary production, and the ecological efficiency. An interesting occurrence since Steele's paper was written is that the catches from the North Sea, which except for periods of war, had remained remarkably constant, especially of demersal fish, have suddenly and rapidly increased. Demersal catches did not vary by more than 350 – 500 thousand tons between 1910 and 1960, with the exception of the periods of the two world wars, but reached over 1 million tons in 1966; pelagic catches increased from between 400 thousand and 800 thousand tons to 1·8 million tons in 1966. Some of these changes, particularly in the pelagic catch, are due to changes in the nature of the fishery, but the changes in demersal catches are mainly due to real changes in the total production of fish (Gulland, 1970).

The primary production does not seem to have changed, so that apparently the ecological efficiency has changed. Steele (1965) discusses changes in the growth of herring that occurred rather earlier, and suggests that a change in the species composition of the zooplankton (from *Temora* and *Pseudocalanus* to the large *Calanus*) has increased the efficiency with which the herring can utilize the food supply. This seems reasonable, and a similar change might have occurred among the demersal stocks, though this has yet to be demonstrated. The interim conclusion is that while the general level of the abundance and production at each trophic level is determined by the primary production, a significant departure from the general level is possible, and probably depends greatly on the detailed species composition at each level, as well as on other environmental factors.

This analysis suggests that food is the ultimate limiting factor for any population, but other evidence shows that such control does not act directly on the exploited part of the population. The food supply presumably acts principally on the growth of the individual fish, but observed changes in growth among fish in exploited stocks have been small. Certainly any such changes have been much less than would be required to maintain the stocks at around their unfished levels in the face of heavy exploitation by man, which often amounts to increasing the mortality rate among the older fish by several-fold. Several studies (Beverton and Holt, 1957; Ricker, 1958) have led to the conclusion that the effective mechanism for maintaining the stock at a more or less stable level lies within the stock/recruitment relation, and that an appropriate shape of the stock/recruit curve (a fairly flat one, or one with a maximum) can result in a very stable population which will not be changed much even under heavy fishing.

There appears to be some discrepancy between these conclusions and the conclusions here that food is important, but further study may well resolve this discrepancy; for instance the food chain argument deals with all the species at one trophic level together; also the pre-recruit stage covers a large segment of a fish's life, not only the stages of eggs and small larvae, during which a range of factors can act on the mortality and growth of the small fish.

VI. Interaction between Species

The previous sections have been concerned mainly with stocks of only one species. No stock exists in isolation, and the fisheries biologist should, like any ecologist, consider all elements of the ecosystem. A full understanding of the factors controlling the population dynamics of one species, say the herring in the North Sea, must include a study of its food (small zooplankton, especially *Calanus* spp.), its competitors for food (e.g. sandeels (*Ammodytes* spp.), or mackerel (*Scomber scombrus*)) and its predators (e.g. cod). Typically, as in the case of the herring, some, but not all, of the species interacting with a commercially valuable fish stock are themselves the object of a commercial fishery; in studying these species the fishery biologist can retain his great advantage of having available the great sampling power of the commercial fleet to do much of the field work for him.

Multi-species studies have, therefore, been mainly confined to situations where both species concerned have been of commercial interest, partly because in these situations the acquisition of scientific data is easier. Another reason is that if, as is likely, separate groups of fishermen exploit the two species, complaints will almost certainly be made by one (if not both) groups against the activities of the other, and fisheries scientists will be called in to assist in the resolution of the conflict. Two types of conflict can be distinguished—when one species is the prey of another, and when the two compete at the same trophic level.

A. FISHING ON PREDATOR AND PREY

A majority of the most valuable fish in the sea are large predators, such as tuna, salmon, cod, halibut. These have typically been the first interest of the fisheries as they developed in an area. Thus tuna are still the only group of fish exploited in the open ocean; halibut and salmon accounted for most of the catch in the north-east Pacific (by weight, but still more by value) until the development in the last few years of trawling for small flounders (*Limanda aspera*), ocean perch (*Sebastodes* spp.) and other fish by the long-range fleets from U.S.S.R. and Japan (Ketchen, 1968). In these areas the interests of the established fisheries have predominated, often excessively, so that the developments of new fisheries, e.g. for herring, have been discouraged because of alleged damage to the salmon.

Recently, in parallel with the expansion of industrialized high-seas fishing into new areas—the South Atlantic, south-east Pacific, etc.—fishing has extended into a greater range of species. This tendency has been strengthened by the increasing proportion of the world's catch which is specifically intended for reduction to meal and oil (mainly for animal feed). This proportion has risen from around 15% in 1958 to 30% in 1965. The fish meal industry depends on a large supply of cheap raw material, so that the species concerned must be very common, or at least occur in dense shoals, and thus are likely to be the food of the larger species supporting earlier commercial fisheries.

These fish meal fisheries have been criticized for misuse of the resources at a time when many people in the world are suffering from protein deficiency. In fact, the difference in price means that a herring fisherman would far rather sell his fish for human consumption than for fish meal, but the hungry man in India or Africa cannot afford to pay enough for fish to cover the costs of preservation and transport of fish to his village from the fishing grounds. Fish are sold for meal and oil only when there is no better market, and since there is no better market for Peruvian anchovy, the present choice is between fishing for meal (with an end-product of cheap chickens), or not fishing them at all. Also, the conversion of anchovy into a form more attractive to the consumer is more efficiently done by the carefully selected chicken to which the fish meal is fed, than by the bonito which eats the anchovy in the sea.

The rise of these meal and oil fisheries has caused concern among the older-established fisheries. Their anxiety is chiefly that the reduction of the food available to the predators will reduce their abundance, or at least drastically affect their movements, so reducing the effectiveness of the fishery on them. In California the concern (mainly of the sport fishermen) over the possible effect of fishing for anchovy (*Engraulis mordax*) has until very recently prevented the development of an industrial fishery for anchovy as the sardine fishery has collapsed. Similar but less effective objections were raised against the anchovy (*E. ringens*) fishery in Peru by the guano interests, fearful of the effect on the birds.

Commonly the predators and the fishery both take a range of species. If so, and especially when the fishery is more selective than the predator, then the effect on the predator is likely to be slight. Thus in the North Sea, sandeels (*Ammodytes* spp.) are at some times and places the major item of diet for cod, but at other times a wide range of other foods are eaten, most of which are not directly exploited by man. Thus even if the recent fishery for sandeels should seriously reduce the stock, as some cod fishermen maintain, there will still remain plenty of alternative foods for cod. While more needs to be known concerning the relation of growth to food supply, it seems unlikely that a reduced supply of just one item in a range of alternatives would have much influence on growth. In other fisheries the predators and the fishery take a similar range of species. Thus off Southern Africa the pelagic fisheries (mainly for meal, but partly for canning) and the birds all catch, in varying proportions, the same species of small shoaling fish—sardines (*Sardinops ocellata*), maasbanker (*Trachurus trachurus*) and anchovy (*Engraulis japanicus*) (see Table I from Davies, 1958 and Matthews, 1961). Information is available concerning the feeding of the birds on the two main fishing grounds—off Walvis Bay, and near Cape Town, about 600 miles further south. In these areas the birds and the shoal fisheries have somewhat similar preferences. Off south-west Africa the choice is limited, and both birds and man rely overwhelmingly on pilchard, while off South Africa there is a wider choice. Heavy fishing on the pilchard stocks would certainly reduce the food supply

TABLE I

Percentage composition of fish consumption by birds (excluding unrecognizable fish remains) and of the S. African catch

	Gannet (*Moris capensis*)		Cormorant (*Phalacrocorax capensis*)		Penguin *Spheniscus demersus*		Man 1951–6	Man 1967	Man 1967
	(*a*)	(*b*)	(*a*)	(*b*)	(*a*)	(*b*)	(*a*)	(*a*)	(*b*)
Pilchard	61	87	58	87	52	94	75	38	97
Maasbanker	18	3	9	9	2	1	43	2	..
Anchovy	17	..	18	1	41	..	..	41	3
Others	4	10	14	3	5	6	..	19	..

(*a*) off South Africa (St Helena Bay). (*b*) off South West Africa (Walvis Bay).

to the Walvis Bay birds (though the extent to which this would affect the bird population is a separate problem), and the same is probably true off South Africa, in that heavy fishing is likely to affect all the main food supplies of the birds.

Complaints by fishermen against predators are also made when the predator is not itself the object of a fishery. Considerable attention has been paid in Scotland to the damage done to the salmon fishery by grey seals (*Halichoerus grypus*) (Rae, 1960). This attention has been far greater than for most such disputes because of the great interest in salmon on one side, and the emotional attachment to seals (especially young seals) among the general public. The exact effects of seals on the fisheries are far from clear, but certain facts are not argued; from a rather low level, though not as low as sometimes stated, which had been brought about by indiscriminate killing, the abundance of grey seals is now increasing following protection at the breeding colonies; seals will take fish from fishing gear and round Scotland salmon in salmon traps are more vulnerable than most other fish which are caught in trawls and seines; seals eat fish and in stomach samples, mainly taken near salmon nets, salmon frequently occur.

It is much less clear how much is eaten by seals, what proportion of the natural food of seals away from salmon nets is salmon, and what is the effect of this feeding on the commercial fisheries. Neither is it clear to what extent the direct damage to nets, and to salmon in the nets, can be reduced by changes in fishing practice. Changes in net design can reduce damage (Shearer, 1962), but changes in position—in the rivers, rather than along the coast—might be even more effective. Another, and perhaps stronger reason for fishing in the rivers, is the experience with Pacific salmon on the west coast of America, where there is a strong belief in the necessity for most fishing to be carried out in or close to the rivers so that the separate stocks in each river can be separately managed—some can be only lightly fished, and others can be much more heavily fished. At present the information necessary for such detailed rational management—especially the relation between the numbers of spawners and subsequent smolt production and hence the optimum number of spawners to be allowed to pass upstream—is not available for most Scottish rivers. This fact has made more difficult the resolution of other salmon problems, such as the effect of the newly developed fishery for salmon, some coming from Canadian and European rivers, at West Greenland (ICES, 1967) but it does not invalidate the ultimate desirability of river-by-river management.

Rae (1962) estimated that grey seals each ate 15 lb (nearly 7 kg) of fish per day and, therefore, including similar quantities eaten by common seal (*Phoca vitulina*), a population of 20 000 grey seals ate a total of 80 000 tons per year. These figures for the consumption per seal and the numbers of seals are admitted to be rather rough, but the range of error probably does not exceed a factor of 2 either way. There is much more uncertainty over the proportion eaten of different species (for instance the 20% salmon or sea trout found by Lockie (1962) in one area would, if applied generally, imply the clearly impossible figure of 16 000 tons of salmon or trout) and over the effect of catching x tons of salmon, or y tons of cod, on the salmon or cod fisheries.

If the y tons were a random sample of the cod population, then the reduction of the cod catch could be readily calculated in terms of the numbers eaten (n), the numbers of these that would have been caught if not eaten by seals $\left(n \times \frac{F}{F+M}\right)$ and their expected average weight. Seals almost certainly do not catch a random sample of the population, especially among the larger and more active species, but take the weaker, sick or more helpless animals—for instance the salmon that are caught in traps. For such animals the calculations above cannot be directly applied, or rather may be applied with the proviso that the values of F and M, and especially M, are not those for the populations as a whole. Thus, it may well be that M owing to disease, etc., is exceptionally high among the fish likely to be preyed on by seals, so that the estimated loss in catch would be much reduced.

Similar considerations can be applied to other predators in order to provide estimates of the damage they do to fisheries and of the degree to which a reduction of predators can be followed by increased catches. Predators which actively pursue individual prey are likely to be selective, and have an effect which is not easy to estimate, nor can their consumption be replaced in all respects by the less selective catches by man. Other predators may not be so selective; at the end of the scale are the baleen whales, straining out astronomical numbers of euphausids at a mouthful. These are clearly not selective and do not select out individuals, though there is probably concentration on the swarms which occur commonly on the surface in Antarctic summer. These are the same swarms of krill (*Euphausia superba*) that would be the objective of direct human exploitation. In this system one may suppose that consumption by whales can be replaced by human exploitation without changing any other elements in the system. A lower limit to the possible harvest from krill is provided by the reduction in consumption by whales following the great reduction in whale stocks by fishing. This comes to around 50 million tons. For comparison, the present world marine fish harvest is also around 50 million tons, and the interest in the possibility of harvesting krill is quite large, the Russians in particular having sent several expeditions to the Antarctic (Burukovskii, 1965).

B. COMPETITION

The most interesting type of interaction between species is that between species occupying similar positions in the food chain, to which the over-used word competition is most often applied. Some of the most drastic changes in the abundance of important food fishes, e.g. the replacement of the sardine by anchovy off California (Murphy, 1966), or of herring by pilchard (sardine, *Sardina pilchardus*) in the English Channel (Cushing, 1966b), have been due to this form of competition, with or without the assistance of fishermen.

Heavy fishing is often supposed to cause the decline of the preferred species at the expense of other less valuable species, e.g. the decline of hake (*Merluccius* spp.) off North-West Africa, and the apparent replacement by cephalopods and other organisms. This is usually very difficult to demonstrate. The species composition in commercial catches indeed often changes, and the percentage of the most valuable species declines as the fishery expands and the total catch increases. These changes are often due to changes in the practice of the fishermen—a small shift of fishing grounds, or differences in rigging the net, can make big alterations in the composition of the catch—made in order to maintain the total catch as the stocks of the initially preferred species decline. Often the less valuable species were discarded when catches were high; when English trawlers first went to the Barents Sea around 1910 only plaice were kept and cod were discarded, and now cod are by far the most important species.

Conclusive evidence of changes in species composition other than catastrophic ones can come only from repeated surveys with research vessels. Because of the expense and time of such surveys they are not common, and rarely cover the whole relevant period, before, during and after the development of a fishery. Sometimes such surveys have shown that changes have followed the simple expectations; thus it appears that in the Bering Sea, where the King crab (*Paralithodes camtschatica*) has been heavily fished for some years, there has been a real (as well as relative) increase in the very similar tanner crabs (*Chionoecetes* spp.) (Alverson, personal communication). In the Gulf of Thailand there have been considerable changes in the species composition (Tiews *et al.*, 1967), but not all these changes are immediately explicable in terms of fishing. The increase in squids that the authors observed might be expected, since squid are not particularly valuable, and are not very vulnerable to the bottom trawls used in the fishery. Equally the decrease in sciaenids might be expected, but snappers (*Lutjanus* spp.) being large and valuable species might also be expected to decrease, but in fact they increased in most of the areas studied in the Gulf.

When changes in the balance of species have occurred following exploitation, two situations can be distinguished—where the non-fishing parameters (growth, recruitment, etc.) of the individual populations have not changed, and where a parameter of one population has been changed by events in another population.

A clear example of the first situation seems to have been the cod and redfish (*Sebastes* spp.) fisheries at West Greenland. The cod fishery, especially by hook and line, is long established, but, more recently, a trawl fishery for redfish was developed by the Germans, for whom redfish is more valuable. Initially, the catch per day of redfish by German trawlers was very high, while few cod were caught. Later, as redfish catches declined cod catches grew and German trawlers in 1966 took as much cod as redfish. Part of this change in proportion of the two species in the catches is due to a change in tactics by the German trawlers, which now concentrate less on the best redfish grounds. This tactical change can be detected and eliminated by using detailed statistics on the catches of individual trawlers, keeping separate the records of those that fished wholly or mainly for cod, those that fished for redfish and those with mixed catches. These detailed statistics show that the redfish have declined more than the cod (but not as much as the catch per unit effort of the German trawl fleet as a whole); this decline can be explained by the slower growth and greater average age of the redfish, so that comparable fishing mortalities on the two stocks will cause a much greater decline in the redfish than will be caused in the cod.

The clearest example of a fishery in which competition has altered the parameters of a population is the sardine fishery off California. The events of this fishery have been clearly described by Murphy (1966). In brief, the total catches increased fairly steadily from 1920 to a peak of nearly 800 000 tons in 1936/7, thereafter declining almost equally steadily to virtually zero in 1967. During the period of most intense fishing in the 1930's the fishing mortality was high—about twice the natural mortality, estimated as around 0·40. Since 1952, while fishing has remained fairly high, there appears to have been an increase in natural mortality to 0·80. The reason for this is not clear, but appears to be connected with a change in population structure to include a greater proportion of a southern race of smaller animals more exposed to predators.

The heavy fishing, and increased natural mortality, can account for only a comparatively small decrease in stock, and a more critical factor has been the decline in recruitment. Murphy found that this could be explained by the use of two separate stock-recruitment relations; one for the period 1932–1948, and one for the period 1949–1957. Within each period there is considerable variation—as in such relations for most fish stocks—but a reasonable fit is given by the Ricker function.

The decline in the sardine has coincided with a marked rise in the abundance of anchovy (*Engraulis mordax*), which until recently has not been fished commercially to any extent. The change in the relative abundance of the two species is shown most conclusively in the catches of larvae in plankton surveys (Murphy, 1966, Table XIX). The food requirements, for both adults and young, are similar for the two species, so that the most economical explanation of the events is, in descriptive terms, that the anchovy increased because the sardine was reduced by fishing, and the presence of abundant anchovies prevented any recovery in the sardine stock.

The explanation can be made in quantitative terms if it is assumed (because of the similarity in food requirements) that the survival of young sardines from egg to recruitment is a function of the combined spawning stock of anchovy plus sardines, and not only of the sardine stock. That is, in terms of the mean natural mortality of the young sardines

$$M_t(\text{sardine}) = a_t + b_t E \text{ (sardine and anchovy)}$$

or in Ricker's formulation

$$R/P = e^{(P_r - P_s \cdot P_A)P_m^{-1}}$$

where P_s = spawning stock of sardine
and P_A = spawning stock of anchovy

or
$$\log R/P = P_m^{-1}(P_r - P_s - P_A).$$

Therefore, if log R/P were plotted against the combined spawning biomass (rather than biomass of sardines only, as in Fig. 11 of Murphy, 1966), the result should be a straight line to which the data for the whole period of data, 1932–1957, should fit.

This plot is shown in Fig. 19. The points for 1932–1948 are shown by open circles and for 1949–1957 by closed circles. The sardine spawning potential and survival have been obtained from Murphy's Table XVIII, and the anchovy spawning biomass estimated from his Fig. 17, assuming a constant population of *ca.* 1·3 million tons before 1940. The estimation of anchovy biomass undoubtedly adds yet further variation to an already variable relation, but it does appear that the same line describes the relation for the entire period, or at least that there is no clear division between the points before and after 1948. The explanation would be much more complete if a similar relation could be shown for anchovy, but the essential data on adult stock and recruitment are missing. One would expect these data to show increased survival in the 1940's and

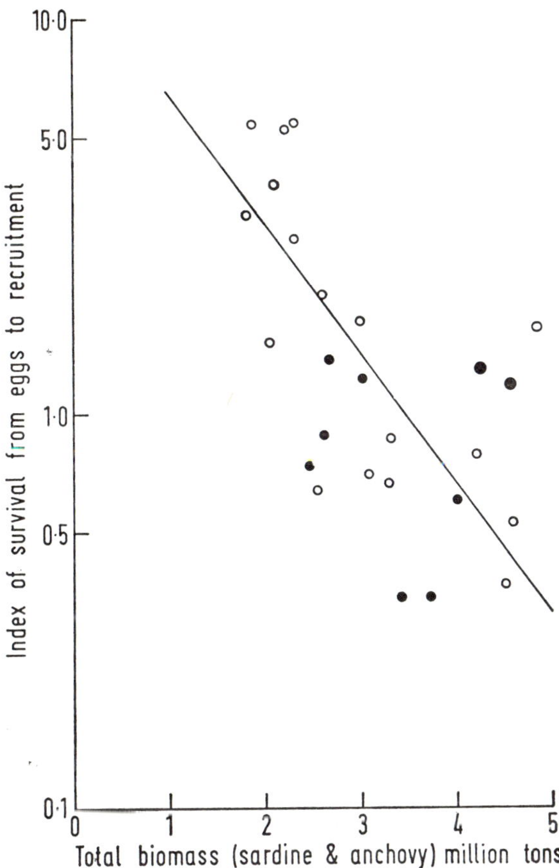

Fig. 19. Survival of Californian sardine from eggs to recruitment, as a function of the combined biomass of adult sardines and anchovies, distinguishing the period 1932–1948 (open circles), and 1949–1957 (closed circles).

early '50's, as the total stock of pelagic fish decreased, and most recently survival similar to that of the early period, but at a much higher level of adult stock, and hence of recruitment.

A further note to the interaction of anchovy and sardines is given by Soutar (1967). He studied the fish scales in bottom deposits off the Californian coast. Three main species have been identified, hake (*Merluccius productus*), anchovy and sardine. The records show that the predominance of the sardine in the 1930's is not typical of the period covered by the deposits, estimated at 1000 years—a total 1430 hake scales, 605 anchovy and 160 sardine scales were counted, though the relation of number of scales to population numbers or population biomass is not known. Analysis of the data showed that the sardine deposits were more aggregated than anchovy or hake (a group of scales was found between the surface and 8 cm deep, and a particularly marked group around 50 cm deep). There is, however, no suggestion that there is a mutual exclusion between sardine and anchovy, as the earlier analysis seems to imply (Table II).

Table II

Occurrence of sardine and anchovy scales in bottom deposits off California (number of layers)

Anchovy scales	Sardine scales		
	Above median	Below median	Total
Above median	20	23	43
Below median	17	18	35
Total	37	41	78

This table, showing the number of layers in which the number of anchovy or sardine scales are above or below the median number shows that these are distributed almost precisely in accordance with the expectations if the two species were independent. It may be that variations in the basic productivity of the area (which might include variations in the position of the upwelling area) tend to produce a positive association, which counteracts and conceals the negative association resulting from competition.

VII. Discussion

Like any other branch of science, fishery science is changing. Past research has been mainly concerned with the larger individuals (the exploited part of the stock) of fish of major commercial importance. There now exist, as described in earlier sections, theoretical frameworks which appear reasonably adequate for studying events (especially the impact of fishing) confined to the exploited phase of the fish's life.

In a few stocks there also exist good estimates of the recruitment, growth and natural mortality during this phase, and the immediate effects of various patterns of fishing, or of other changes within the exploited phase, can be readily deduced. For many more stocks these parameters still need to be estimated, using established methods. The theoretical framework and the methods of estimating parameters do, however, need to be developed to take into account all elements of the ecosystem.

Developments into some of these aspects are particularly likely in the near future with the increased attention now being paid to all aspects of ocean science, some of which is directly inspired by interest in fishing.

This increased activity will greatly increase the flow of information to the fishery scientist. It will also change the balance among the different sources of information—the commercial fishing vessel may cease to be the major, and sometimes the only, source of information—and hence, ultimately, the techniques used. The biggest change is likely to be increased information on the weather and climate in the oceans—more strictly on the variations in the ocean currents and in the physical conditions in the sea such as temperature. Ten years ago, such information was limited to very infrequent spot observations by research vessels, plus rather more frequent but cruder data from merchant ships and fishing vessels. Now, with the realization of the importance of sea conditions to the weather over the land a whole range of new techniques has come into being. Satellites are providing excellent data on surface conditions, buoys are being developed which can be moored in the ocean for weeks at a time recording automatically and continuously a whole range of such items as temperature, current speed and direction, at the surface or at pre-selected depths, and research ships can make similar observations continuously while steaming without needing to stop.

This better knowledge of the physical environment will obviously be useful to fisheries and fisheries research, espcecially in the application to improved forecasting of the distribution and availability of fish stocks, and in the removal of some of the variation that obscures many of the important biological relations, e.g. between adult stock and subsequent recruitment.

Good knowledge merely of the exploited stocks and of the physical environment is, however, not sufficient to understand what is controlling the fish population, or to provide all the required advice to the fishing industry especially concerning the needs for conservation and management. Previous sections have outlined some of the fields in which much better research is required, for example events in the first few months or weeks of life, leading hopefully to better understanding of the stock/recruitment relation, and thus of the mechanisms which tend to maintain many fish stocks at a more or less constant level. Other fields in which research is as yet barely beginning include the evolutionary and selective significance of the various characteristics of fishes, especially those which are markedly different from land animals. For instance, why do fish, and indeed marine animals generally have such high fecundity, and what selective advantage, if any, does the cod, which produces perhaps a million eggs per year, have over the dogfish—about the same size and age, with not very different food—which only produce half a dozen young every other year?

At the moment fishery scientists are only just beginning to pose these and similar questions in other fields; when they are answered, it may be possible to claim that there is a fair understanding of what is happening to fish in the sea and to hope that there will be a more efficient and less wasteful exploitation of the fishery resources of the world.

Acknowledgements

My thanks are due to D. H. Cushing, G. L. Kesteven and S. J. Holt for critical readings of the draft of this paper, and to H. E. Jenner for preparing the figures, to Miss D. Spencer for typing the manuscript, and to Miss G. Soave for assistance with the bibliography. Also to the following for permission to reproduce the figures mentioned: H.M. Stationery Office (Fig. 8), Inter-American Tropical Tuna Commission (Fig. 7), Blackwell Publications Ltd. (Figs. 15, 16), and H. W. Graham (Fig. 2).

References

Ahlstrom, E. H. (1954). *Fishery Bull. Fish Wildl. Serv. U.S.* **56**, 83–140. Distribution and abundance of egg and larval populations of the Pacific sardine.

Ahlstrom, E. H. (1965). *Spec. Publs. int. Commn NW. Atlant. Fish.* 53–74. A review of the effects of the environment of the Pacific sardine.

Allee, W. C., Finkel, A. J., and Hoskins, W. H. (1940). *J. exp. Zool.* **84**, 417–443. The growth of goldfish in homotypically conditioned water: population study in mass physiology.

Andrewartha, H. G. and Birch, L. C. (1954). "The distribution and abundance of animals." University of Chicago Press, 782 pp.

Anon. (1958). Fish Stock Record 1957, Ministry of Agriculture, Fisheries and Food, Fisheries Laboratory, Lowestoft (mimeo).

Anon. (1966). ICES Herring Committee and Gadoid Fish Committee. (H:23) (mimeo). Preliminary report of the joint international O-group fish survey on the Barents Sea and adjacent waters. August/Sept. 1966.

Applegate, V. C. (1950). *Spec. scient. Rep. U.S. Fish Wildl. Serv. (Fish.)* (55), Natural history of the sea lamprey, *Petromyzon marinus* in Michigan.

Backiel, T. and Le Cren, E. D. (1967). *In* "The biological basis of freshwater fish production" (S. D. Gerking, ed.) Blackwell, Oxford, pp. 261–294. Some density relationships for fish population parameters.

Baranenkova, A. S. (1960). *In* "Soviet fishery investigations in North European seas." VNIRO/PINRO, pp. 267–275. The results of young cod and haddock surveys in the Barents Sea in the period 1946–50 (in Russian).

Baranov, F. I. (1918). *Izv. nauchno.-issled. ikthiol. Inst.* **1**, 81–128. On the question of the biological basis of fisheries (in Russian).

Bertalanffy, L. von (1938). *Hum. Biol.* **10**, 181–223. A quantitative theory of organic growth.

Bertalanffy, L. von (1957). *Q. Rev. Biol.* **32**, 217–231. Quantitative laws in metabolism and growth.

Bertelsen, E. and Popp Madsen, K. (1953). *Ann. biol. Copenh.* **9**, 179–180. Young herring from the Bloden ground area.

Beverton, R. J. H. (1953). *J. Cons. perm. int. Explor. Mer* **19**, 56–68. Some observations on the principles of fishery regulation.

Beverton, R. J. H. (1962). *In* "The exploitation of natural animal populations" (E. D. Le Cren and M. W. Holdgate, eds.) Blackwell, Oxford, pp. 242–259. Long-term dynamics of certain North Sea fish populations.

Beverton, R. J. H. and Hodder, V. M. (eds) (1962). *A. Proc. int. Commn NW. Atlant. Fish.* **11** (Suppl.). Report of the working group of scientists on fishery assessment in relation to regulation problems.

Beverton, R. J. H. and Holt, S. J. (1956). *In* "Sea fisheries: their investigation in the United Kingdom" (Michael Graham, ed.) Edward Arnold Ltd., London, pp. 372–441. The theory of fishing.

Beverton, R. J. H. and Holt, S. J. (1957). *Fishery Invest. Lond.* (2), 533 pp. On the dynamics of exploited fish populations.

Brody, S. (1945). "Bioenergetics and growth." Reinhold, New York.

Brown, M. E. (1946). *J. exp. Biol.* **22**, 118–129. The growth of brown trout (*Salmo trutta* L.) 1. Factors influencing the growth of trout fry.

Burukovskii, R. N. (ed.) (1965). "Antarctic krill. Biology and industry." Kaliningrad.

Carruthers, J. N. (1938). *Rapp. P.-v. Reun. Cons. perm. int. Explor. Mer* **107**, 10–15. Fluctuations in the herrings of the East Anglian autumn fishery, the yield of the Ostend spent herring fishery and the haddock of the North Sea in the light of relevant wind conditions.

Carruthers, J. N., Lawford, A. L. and Volay, V. F. C. (1951). *Kieler Meeresforsch.* **8**, 5–15. Fishery hydrography: brood strength fluctuations in various North Sea fish with suggested methods of predictions.

Chapman, D. G. (1964). *Rep. int. Commn Whal.* **14**, 32–106. Reports of the Committee of Three Scientists on the Special Scientific Investigation of the Antarctic Whale Stocks.

de Ciechomski, J. D. (1967). *Rep. Calif. co-op. ocean. Fish. Invest.* (11), 72–81. Investigations of food and feeding habits of larvae and juveniles of the Argentine anchovy.

Clark, F. N. and Marr, J. C. (1955). *Prog. Rep. Calif. co-op. ocean. Fish. Invest.*, I July 1953–31 March 1955, pp. 11–48. Population dynamics of the Pacific sardine.

Corlett, J. (1965). *Spec. Publs. int. Commn NW. Atlant. Fish.* (6). 373–378. Wind, currents, plankton and the year-class strength of cod in the western Barents Sea.

Cushing, D. H. (1966a). "The Arctic cod. A study of research into the British trawl fisheries." Pergamon Press, London.

Cushing, D. H. (1966b). *Biol. Rev.* **41**, 221–258. Biological and hydrographic changes in British Seas during the last thirty years.

Cushing, D. H. (1968). "Fisheries Biology. A study in population dynamics." University of Wisconsin Press, 200 pp.

Davies, D. H. (1958). *Investl Rep. Div. Fish. Un. S. Afr.* (31). The South African pilchard and maasbanker. The predation of sea-birds in the commercial fishery.

Dickie, L. M. and McCracken, F. D. (1955). *J. Fish. Res. Bd Can.*, **12**, 187–209. Isopleth diagrams to predict equilibrium yields of a small flounder fishery.

Doi, T. (1962). *Bull. Tokai Reg. Fish. Lab.* (32), 49–118. The predator-prey and competitive relationships among fishes caught in waters adjacent to Japan (in Japanese).

Graham, H. W. (1966). *Red Bk. int. Commn NW. Atlant. Fish.* 1966, 106–121. United States Research report, 1965.

Graham, H. W. (1967). *Red Bk int. Commn NW. Atlant. Fish.* 1967, 136–151. United States Research report, 1966.

Graham, M. (1939). *Rapp. P.-v. Réun. Cons. perm. int. Explor. Mer.* **110**, 15–20. The sigmoid curve and the over-fishing problem.

Grygierek, E. (1962). *Riczn. Nauk roln.* (*B*) (2), 189–210. The influence of increasing carp fry population on crustacean plankton (in Polish, Eng. summary).

Grygierek, E. (1965). *Riczn. Nauk roln.* (*B*) (2), 147–168. The effect of fish on crustacean plankton (in Polish, Eng. summary).

Gulland, J. A. (1953). *J. Cons. perm. int. Explor. Mer* **18**, 351–353. Letter to the editor. Correlations on fisheries hydrography.

Gulland, J. A. (1955). *Fishery Invest. Lond.* (2) **18**. On the estimation of growth and mortality in commercial fish populations.

Gulland, J. A. (1956). *Fishery Invest. Lond.* (2) **20**. On the fishing effort in English demersal fisheries.

Gulland, J. A. (1961). *Fishery Invest. Lond.* (2) **23**. Fishing and the stocks of fish at Iceland.

Gulland, J. A. (1962). *In* "The exploitation of natural animal populations" (E. D. Le Cren and M. W. Holdgate, eds.) Blackwell, Oxford, pp. 204–217. Application of mathemetical models to fish populations.

Gulland, J. A. (Ed.), (1964). *Rapp. P.-v. Reun. Cons. perm. int. Explor. Mer* **155,** Contributions to symposium 1963 on the measurement of abundance of fish stocks.

Gulland, J. A. (1965). *Spec. Publs. int. Commn NW. Atlant. Fish.* (6), 363–372. Survival of the youngest stages of fish, and its relation to year-class strength.

Gulland, J. A. (1966). *J. Cons. perm. int. Explor. Mer* **30**, 308–315. Effects of regulations on Antarctic whale catches.

Gulland, J. A. (1968a). *J. Cons. perm. int. Explor. Mer* **32**, 256–261. Concept of the marginal yield from exploited fish stocks.

Gulland, J. A. (1968b). *FAO Fish. tech. Pap.* (70). Concept of the maximum sustainable yield and fishery management.

Gulland, J. A. (1968c). *FAO Fish. tech. Pap.* (40), Rev. 2. Manual of methods for fish stock assessment. 1. Fish population analyses.

Gulland, J. A. (1968d). *Boln Inst. Mar Peru* **1**, 305–346. Informe sobre la dinámica de la población de Anchoveta peruana (Transl.).

Gulland, J. A. and Carroz, J. E. (1968). *Adv. mar. biol.* **6**, 1–71. Management of fishery resources.

Gulland, J. A. (1970). Food Chain Studies and some problems in world fisheries. In J. Steele (ed.) "Marine food chains." Oliver and Boyd, Edinburgh.

Herman, F., Hansen, P. M. and Horsted, S. A. (1965). *Spec. Publs. int. Commn NW. Atlant. Fish.* (6), 389–396. The effect of temperature and currents on the distribution and survival of cod larvae at West Greenland.

Hickling, C. F. (1934). "The hake and the hake fishery." Edward Arnold, London.

Hjort, J. (1914). *Rapp. P.-v. Reun. Cons. perm. int. Explor. Mer* **20**. Fluctuations in the great fisheries of northern Europe viewed in the light of biological research.

Hjort, J. (1926). *J. Cons. perm. int. Explor. Mer* **1**, 5–38. Fluctuation in the year-classes of important food fisheries.

Holden, M. J. (1960). *J. Cons. perm. int. Explor. Mer* **26**, 68–72. Evidence of cod (*Gadus morhua* L.) migrations from the Norway coast to the Faroes Islands.

Hrbacek, J., Dvorakova, M., Korinele, V. and Prochazkova, L. (1961). *Verh. int. Verein. theor. angew. Limnol.* **14**, 192–195. Demonstrations on the effect of the fish stocks on the species composition of zooplankton.

ICES (1966a). *Co-op. Res. Rep. int. Coun. Explor. Sea (B)* 1966, 15–32. Report of the Liaison Committee of ICES to the North-East Atlantic Fisheries Commission, 1965. Annex I. Report of the Meeting on the Arctic Fisheries Working Group in Hamburg, 18–23 January 1965.

ICES (1966b). *Co-op. Res. Rep. int. Coun. Explor. Sea (B)* 1966, 33–79. Report of the Liaison Committee of ICES to the North-East Atlantic Fisheries Commission, 1965. Annex II. Report of the Assessment Group in herring and herring fisheries in the north-eastern Atlantic.

ICES (1967). *Co-op. Res. Rep. int. Coun. Explor. Sea (A)* (8). Report of the ICES/ICNAF Joint Working Party on North Atlantic Salmon, 1966.

Ivlev, V. S. (1961). Experimental Ecology of the feeding of fishes. New Haven and London. Yale Univ. Press, 302 pp.

Jones, R. (1966). ICES, Gadoid Fish. Committee, 1966, Doc. (G20). Post-war changes in the North Sea stock of haddock.

Ketchen, K. S. (1968). *Trans. Am. Fish. Soc.* **97**, 82–88. Trends in marine fisheries along the Pacific Coast of North America.

Koblents-Mishke, O. I. (1965). *Okeanologiia* **5**, 325–337. Value of primary productivity in the Pacific Ocean (in Russian).

Larrañeta, M. G. (1967). "Ecologia marina." Fundación la salle, Caracas.

Le Cren, E. D. (1965). *Mitt. int. Verein. theor. agnew. Limnol.* **13**, 88–105. Some factors regulating the size of populations of freshwater fish.

Leslie, P. H. (1957). *Biometrika* **44**, 314–327. An analysis of the data for some experiments carried out by Gause with populations of the protozoa *Paramecium aurelia* and *P. caudatum.*

Lockie, J. D. (1962). *In* "The exploitation of natural populations" (E. D. Le Cren and M. W. Holdgate, eds.) Blackwell, Oxford, pp. 316–325. Grey seals as competitors with man for salmon.

Lumby, J. R. and Atkinson, G. T. (1929). *J. Cons. perm. int. Explor. Mer* **4**, 309–322. On the unusual mortality amongst fish during March and April 1929.

Margetts, A. R. (1949). *Rapp. P.-v. Reun. Cons. perm. int. Explor. Mer* **125**, 72–81. Experimental comparison of fishery capacities of different trawlers and trawls.

Margetts, A. R. (1949a). Experimental comparison of fishing capacities of different trawlers and trawls. *Rapp. Cons. Explor. Mer* **125**, 72–81.

Margetts, A. R. (1949b). Experimental comparison of fishing capacities of Danish seiners and trawls. *Rapp. Cons. Explor. Mer* **125**, 82–90.

Marr, J. C. (1956). *J. Cons. perm. int. Explor. Mer* **21**, 160–170. The "critical period" in the early life history of marine fishes.

Marr, J. C. (1960). *In* Proceedings of the World Scientific Meeting on the Biology of Sardines and Related Species held in Rome, 14–21 September 1959, (H. Rosa, Jr. and G. Murphy, eds.), Rome, FAO, **3**, 667–791. The causes of major variations in the catch of the Pacific sardine *Sardinops caerulea* (Girard).

Matthews, J. P. (1961). *Investl Rep. mar. Res. Lab. S.W. Afr.* (3), 35 pp. The pilchard of south west Africa and the maasbanker (*Trachurus trachurus*) bird predators, 1957–1958.

Menasveta, D. (1968). Paper presented to the CSK Symposium, 29 April–2 May 1968, Honolulu, Mimeo. Potential demersal fish resources of the Sunda Shelf.

Murphy, G. I. (1966). *Proc. Calif. Acad. Sci.* **34**, 84 pp. Population biology of the Pacific sardine (*Sardinops caerulea*).

Nakai, Z., Usami, S., Hatori, S., Honjo, K. and Hayashi, S. (1955). Tokyo, Tokai Regional Fisheries Research Laboratory, Progress report of the cooperative Iwashi resources investigations, April 1949–December 1951.

Nicholson, A. J. (1954). *Aust. J. Zool.* **2**, 9–65. An outline of the dynamics of animal populations.

Nicholson, A. J. (1958). *A. Rev. Ent.* **3**, 107–136. Dynamics of insect populations.

Paloheimo, J. E. and Dickie, L. M. (1966). *J. Fish. Res. Bd Can.* **23**, 1209–1248. Food and growth of fish. 3. Relations among food, body size and growth efficiency.

Paloheimo, J. E. and Dickie, L. M. (1970). *In* "Marine Food Chains" (J. M. Steel, ed.). Oliver and Boyd, Edinburgh, pp. 499–527. Production and food supply.

Parker, R. R. and Larkin, P. A. (1959). *J. Fish. Res. Bd Can.* **16**, 721–745. A concept of growth in fishes.

Parrish, B. B. (1956). *In* "Sea fisheries". (M. Graham, ed.) Edward Arnold, London, pp. 251–331. The cod, haddock and hake.

Paulik, G. and Bayliff, W. K. (1967). *J. Fish. Res. Bd Can.* **24**, 249–259. A generalized computer program for the Ricker model of equilibrium yield per recruitment.

Pella, J. J. (1967). Thesis, University of Washington, 155 pp. "A study of methods to estimate the Schaefer model parameter with special reference to the yellowfin tuna fishery in the eastern tropical Pacific Ocean." Ph.D. Thesis, University of Washington.

Ponomarenko, V. P. (1967). Paper presented to ICES Meeting, 1967, Document (F16). Reasons for changes in the rate of growth and maturation of the Barents Sea Cod.

Rae, B. B. (1960). *Mar. Res.* 1060, 1–39. Seals and Scottish fisheries.

Rae, B. B. (1962). *In* "The exploitation of natural animal populations". (E. D. LeCren and M. W. Holdgate, eds.) Blackwell, Oxford, pp. 305–211. The effect of seal stocks on Scottish marine fisheries.

Raitt, D. S. (1939). *Rapp. P.-v. Reun. Cons. perm. int. Explor. Mer* **110**, 65–79. The rate of mortality of North Sea haddock stock 1919–1938.

Richards, F. J. (1959). A flexible growth function for empirical use. *J. exp. Bot.* **19** (29), 290–300.

Ricker, W. E. (1948). *Indiana Univ. Publs. Sci. Ser.* **15**, 101 pp. Methods of estimating statistics of fish populations.

Ricker, W. E. (1954). *J. Fish. Res. Bd Can.* **11**, 559–623. Stock and recruitment.

Ricker, W. E. (1958). *Bull. Fish. Res. Bd Can.* (119). Handbook of computations for biological statistics of fish populations.

Royce, W. F., Bevan, D. E., Crutchfield, J. A., Paulik, G. J. and Fletcher, R. C. (1963). *Univ. Wash. Publs Fish.* **2**, 1–123. Salmon gear limitations in northern

Washington waters; and economic, biological and legal survey of the salmon resource of northern Puget Sound and Strait of Juan de Fuca.

Russell, E. S. (1942). "The overfishing problem." Cambridge University Press.

Saville, A. (1959). *Mar. Res.* **3**, 23 pp. The planktonic stages of haddock in Scottish waters.

Saville, A. S. (1965). *Spec. Publs. int. Commn NW. Atlant. Fish.* (6), 335–348. Factors controlling dispersal of the pelagic stages of fish and their influence on survival.

Schaefer, M. B. (1954). *Bull. inter-Am. trop. Tuna Commn* **1**, 27–56. Some aspects of the dynamics of populations important to the management of commercial marine fisheries.

Schaefer, M. B. (1957). *Bull. inter-Am. trop. Tuna Commn* **2**, 245–285. A study of the dynamics of yellowfin tuna in the eastern tropical Pacific.

Schaefer, M. B. (1967a). *Bull. inter-Am. trop. Tuna Commn* **12**, 89–136. Fishery dynamics and present status of yellowfin tuna population of the eastern Pacific Ocean.

Schaefer, M. B. (1967b). *Boln Inst. Mar Peru* **1** ,191–303. Dynamics of the fishery for the anchoveta *Engraulis ringens*, off Peru.

Schaefer, M. B. and Beverton, R. J. H. (1963). *In* "The sea" (M. N. Gill, ed.), (Wiley, New York), **Vol. 2**, 464–483. Fishery dynamics—their analysis and interpretations.

Sette, O. (1943). *Fishery Bull. Fish Wildl. Serv. U.S.* **50**, 149–237. Biology of the Atlantic mackerel (*Scomber scombrus*) of North America. Part I. Early life history, including growth, drift and mortality of the egg and larval populations.

Shearer, W. M. (1962). *In* "The exploitation of natural animal populations" (E. D. Le Cren and M. W. Holdgate, eds.) Blackwell, Oxford, pp. 312–315. Seals and salmon nets.

Shelbourne, J. E. (1957). *J. mar. biol. Ass. U.K.* **36**, 539–552. The feeding and condition of plaice larvae in good and bad plankton patches.

Silliman, R. P. (1966). *Fishery Bull. Fish Wildl. Serv. U.S.* **66**, 31–46. Analog computer models of fish populations.

Silliman, R. P. and Gutsell, J. S. (1958). *Fishery Bull. Fish Wildl. Serv. U.S.* **58**, 214–252. Experimental exploitation of fish population.

Sindermann, C. (1958). *Trans. N. Am. Wildl. Conf.* **23**, 349–360. An epizootic in Gulf of St. Lawrence fishes.

Slobodkin, L. B. (1962). *In* "Exploitation of natural animal populations" (E. D. Le Cren and M. W. Holdgate, eds.) Blackwell, Oxford, pp. 223–241. Predation and efficiency in laboratory populations.

Soutar, A. (1967). *Rep. Calif. co-op. ocean. Fish Invest.* (11), 136–139. The accumulation of fish debris in certain California coastal sediments.

Southward, G. M. (1968). *Rep. int. Pacif: Halibut Commn* (47), 70. A simulation of management strategies in the Pacific halibut fishery.

Steele, J. (1965). *Spec. Publs. int. Commn NW. Atlant. Fish.* (6), 463–476. Some problems in the study of marine resources.

Steele, J. (Ed.) (1970). "Marine Food Chains". Oliver and Boyd, Edinburgh.

Storey, M. and Gudger, E. W. (1936). *Ecology* **17**, 640–648. Mortality of fishes due to cold at Sanibel Island, Florida, 1886–1936.

Templeman, W. (1965). *Spec. Publs. int. Commn NW. Atlant. Fish.* (6), 137–148. Mass mortalities of marine fishes in the Newfoundland area presumably due to low temperature.

Tiews, K. (1965). *Arch. Fisch Wiss.* 16 (1): 67–108. Bottom fish resources investigation in the Gulf of Thailand and an outlook on further possibilities to develop the marine fisheries in southeast Asia.

Tiews, K., Sucondhamarn, P. and Isarankura, A. (1967). Bangkok, Marine Fisheries Laboratory, Contribution (8): 39 pp. On the changes in the abundance of demersal fish stocks in the Gulf of Thailand from 1963/1964 to 1966 as a consequence of the trawl fisheries development.

Townsend, C. H. (1935). *Zoologica, N.Y.* **19**, 1–50. The distribution of certain whales as shown by logbook records of American whaleships.

Volterra, V. (1931). "Leçons sur la théorie mathématique de la lutte pour la vie." Gauthier-Villars, Paris.

Walter, E. (1934). *Handb. Binnenfisch. Mittleleur.* **4**, 480–662. Grundlagen der all gemeine fischerielischen Produktionslehre.

Watt, K. E. F. (1968). "Ecology and resource management: A quantitative approach." McGraw-Hill, New York.

Wimpenny, R. S. (1953). "The plaice." The Buckland lectures for 1949. Edward Arnold, London.

Winsor, C. P. (1932a). *Proc. Natn. Acad. Sci. U.S.A.* **18**, 1–8. The Gomperts curve as a growth curve.

Winsor, C. P. (1932b). *J. Wash. Acad. Sci.* **22**, 73–84. A comparison of certain symmetrical growth curves.

Vegetational Distribution, Tree Growth and Crop Success in Relation to Recent Climatic Change

J. R. BRAY

P.O. Box 494, Nelson, New Zealand

I. Introduction

Knowledge of post-Pleistocene climates, and particularly temperatures, has been greatly increased recently by research in both the physical and biological sciences. With more exact delineation of long-term climatic trends, it is possible to evaluate some of the changes in vegetational distribution and plant growth which may be related to these trends. This review will cover mainly the past 2000 years, an interval which includes most historical human observations and measurements. Because of the critical re-evaluation which is being made at present of the techniques and especially of some of the interpretations of pollen analysis, this aspect is not included.

II. Climate

A. post-pleistocene

The predominant temperature trend since the Pleistocene was the extensive warming in the middle to higher latitudes and altitudes which culminated in a plateau of maximum warmth and was followed by a minor decline to the present. This interval of warming accompanied, and largely resulted from, the melting of the Pleistocene mid-latitude ice sheets and the consequent rise in ocean level; the subsequent minor cooling has resulted in an increase in arctic and alpine glaciation and a slight fall in ocean level. Data summarized in Table I indicate that

Table I

Climatic data, 0–10 000 BP

		Temperature °C	Temperature °C
Millennium BP	Sea level m (Schofield, 1964)	40°–90°N latitude (from Dorf, 1960)	central England (from Lamb *et al.*, 1966)
10	−35·8	−1·4	0·0
9	−33·8	−0·6	5·0
8	−18·9	−0·7	9·8
7	−4·8	−0·3	12·1
6	−0·1	+1·7	11·5
5	—1·6	+1·9	10·5
4	5·1	+1·6	9·9
3	2·8	+0·8	10·0
2	1·2	0·0	10·2
1	0·8	0·0	10·1

the temperature optimum was reached around the 7th–5th millennia BP, after which there was still sufficient warmth to result in continued rise in sea level through the 4th millennium BP. A secondary post-Pleistocene temperature pattern, of particular significance for the past two millennia, was characterized by recurrent cold phases, with each phase identified by a period of increased high latitude and high altitude glaciation (Bray, 1970a). This pattern was of lesser magnitude than the predominant pattern and was apparently superimposed upon it, with the recurrent cold phases most evident towards the cooler portions of the predominant pattern, that is, early in the post-Pleistocene and, again, recently.

These two Holocene temperature trends were probably of independent origin, as shown by the appearance of a glacial phase close to the Thermal Maximum and, therefore, were determined by different physical mechanisms (Bray, 1970a). The slope of the predominant trend was obviously greatly influenced by the warming from the melting of the Pleistocene ice. The actual change in energy budget which produced this warming may have been relatively small (Adam, 1969; Kukla, 1969). Movements of the earth's orbit relative to the sun (the Milankovitch hypothesis) have been suggested as the most likely reason for this warming trend. According to a recent summary of this hypothesis by Kukla (1969), changes in the seasonal distribution of incoming solar energy would influence the areal distribution of highly reflective sheets of snow and pack ice, particularly in the middle latitudes and especially in the northern hemisphere. If winter insolation over the northern hemisphere were decreasing, snow and snow-covered pack ice would extend and global cooling would follow. An increase in winter insolation over the northern hemisphere would result in a decrease in the snow sheet with subsequent global warming.

Delineation of the secondary temperature trend, that of recurrent cold phases of short duration, was based on an analysis of post-Pleistocene glaciation (Bray, 1968, 1970a) which showed that while some glacial advance had occurred throughout the period, most of it was confined to four major phases. These phases will be briefly outlined since it is the last one, from around AD 1550–1900, which forms the most conspicuous climatic aberration of the past 2 millennia and for which the most palaeoecologic information is available.

The first glacial phase, the Cochrane–Cockburn (Ragunda?) lasted several hundred years and culminated around 8000–9000 ^{14}C years BP, probably nearer 8000 BP. This phase represents the only glacial pulsation that can be recognized stratigraphically and geomorphically over large areas of the Canadian eastern and central Arctic. It is also the only post-Pleistocene glacial phase with any significant amount of middle latitude low elevation ice advance. The Cochrane–Cockburn is considered to be the final phase of the late Wisconsin and the last period in which the Laurentide ice sheet can be recognized as a whole. The second glacial phase was the world-wide resurgence at the Atlantic/sub-Boreal transition period around 4600–5200 ^{14}C years BP, with most activity around 4600–4800 years. This brief period of cooling occurred near the post-Pleistocene Thermal Optimum, was the least extensive of the four glacial phases and the most recent to be clearly recognized. The next main period of glaciation was during the sub-Atlantic minimum and culminated around 2600–2800 ^{14}C years BP. This phase has been well documented and was apparently of greater extent than the preceding phase, at least in North America, where Bassett and Terasmae (1962) suggest there may have been local snow accumulation in Labrador-Ungava. The most recent glacial phase was the advance of around AD 1550–1900, the "Little Ice Age" (hereafter abbreviated LIA) for which a world-wide chronology has been summarized (Bray, 1968). Glacial advance during this period was on average more extensive than in the sub-Atlantic minimum (Porter and Denton, 1967) and in a few cases ice from higher elevations reached the piedmont or sea level. Conversion of dates for these four main glacial phases from ^{14}C time to true time resulted in corrected dates of around 8250, 5350, 2780 and 225 years BP (Bray, 1970a). These corrected dates substantiated the original estimate of a cycle of around 2600 years which was presumed to be based on recurrent periods of lowered solar activity.

B. RECENT: 2000 BP–PRESENT

1. Temperature—Physical measurements

During the 8th–12th centuries AD, mean temperature for 40–90°N latitude (Dorf, 1960) was the warmest since the Thermal Maximum around 4000 BC, and the period AD 1550–1750 was the coldest since 500 BC. There was a maximum difference of 1·9°C between the coldest century, the 16th, and the warmest century, the 11th, as shown in Table II. The magnitude of the difference between these two extreme periods was least in the equatorial regions and greatest at higher latitudes and altitudes. This is evident from the mean annual temperature trends summarized by Willett (1950), which show differences since 1850 between the coolest (1880) and the warmest (1930) decades that range from less than 1°C in the tropics to greater than 3°C in the latitude 70–80°. Individual stations show even greater rises such as the increase of 4°C from 1912–1920 to 1930–1940 at Spitzbergen, Norway (Lamb, 1965).

The most reliable temperature trend over the past millennia is the Manley–Lamb curve for England (Lamb, 1965) and the mean of the winter mildness/severity index for 50°N latitude for England, Germany and Russia (Lamb, 1966). These data are shown in Table III and indicate a general decline from the maximum in the 12th and 13th centuries to a minimum around 1550–1700 followed by a slow increase to the present.

TABLE II

Climatic data, 0–2000 BP

Century AD	T°C 40–90°N latitude (from Dorf, 1960)	Relative world glaciation (from Porter and Denton, 1967)	No. years Thames frozen (from Lamb, 1959)	No. years good wine harvest Baden (from Lamb, 1959)	Botanical data showing warmth or a warming trend %	No.
20	+0·6	2	—	38	79	19
19	−0·4	18	—	27	16	3
18	−0·6	19	6	40	47	9
17	−0·7	14	8	30	0	0
16	−0·8	9	4	36	25	4
15	−0·6	8	1	32	18	4
14	0·0	9	0	42	17	4
13	+0·6	4	2	44	67	10
12	+0·8	2	2	57	91	10
11	+1·1	1	1	44	90	9
10	+1·0	2	1	46	75	6
9	+0·9	2	—	62	100	7
8	+0·8	2	—	—	67	4
7	+0·7	2	—	—	100	4
6	+0·6	2	—	—	100	3
5	+0·4	1	—	—	100	3
4	+0·3	1	—	—	100	4
3	+0·2	0	—	—	100	3
2	0·0	0	—	—	100	3
1	0·0	1	—	—	100	3

2. Temperature—Geophysical measurements

The temperature patterns outlined by the Dorf and Lamb data in Tables II and III are supported (Table III) by three geophysical indexes, ^{14}C, ^{18}O and solar activity, which are related to temperature. The ^{14}C data are a summary of all values used by Bray (1967). Lower ^{14}C values were considered an index of higher temperatures (Suess, 1965; Bray, 1966a). These data were sampled mainly in western North America and Europe, but because of the efficiency of atmospheric mixing can be considered to represent a general global trend. The ^{14}C data indicate warmer temperatures from AD 0 to 1400, with peak intervals around 150–400 and 800–1250 and a minor regression to cooler temperatures around 600–700. Following the 800–1250 period there was a major regression from 1400 to 1850 with a minimum around 1500–1750 and a subsequent temperature rise to the present.

The ^{18}O values are from the Camp Century, Greenland ice core

TABLE III

Climatic indexes

Period AD	Yearly temp. central England °C (from Lamb, 1965)	Winter mildness/severity index Europe near 50°N (from Lamb, 1966)	δ¹⁴C (‰)	δ¹⁸O (‰) (from Dansgaard *et al.* 1969)	Solar activity index	Precipitation % 1916–1950 mean England and Wales (from Lamb, 1965)	High summer wetness/dryness index Europe near 50°N (from Lamb, 1966)	Drift ice at Iceland weeks/year (from Koch, 1945)	Number recorded maximum glacial advances: Northern hemisphere	Number recorded maximum glacial advances: Southern hemisphere	Relative tree growth inversed and smoothed (stands 9+10 of Fritts, 1965a)
1900–1950	9·4	+44	−0·8	−284	99	99	29·9	4·9	5·5	0·0	−1·02
1850–1900	9·1	+8	0·0	−287	94	97	34·3	17·2	14·0	4·0	+0·27
1800–1850	9·1	−18	+0·4	−294	79	96	34·8	16·3	23·5	3·0	+0·94
1750–1800	9·1	−48	+0·4	−283	109	94	32·3	16·6	15·5	6·5	+0·91
1700–1750	9·2	−16	+1·3	−291	86	96·5	30·1	9·3	25·5	1·5	+0·31
1650–1700	8·7	−121	+1·2	−298	50	92·5	33·5	7·8	3·5	2·5	−1·84
1600–1650	8·8	−84	+0·9	−296	88	93·5	30·9	11·5	7·5	1·0	−2·51
1550–1600	8·8	−36	+0·4	−290	126	93·5	35·7	2·5	4·5	1·0	+0·82
1500–1550	9·3	+5	+1·5	−293	103	97	27·6	0·7	2·0	1·5	+2·58
1450–1500	9·0	−59	+0·8	−289	63	95	34·1	1·6	2·0	0	+1·10
1400–1450	9·1	−106	+0·8	−291	67	95	27·8	0·0	0	0	+1·30
1350–1400	9·5	−32	−0·5	−289	115	97·5	29·0	2·0	0	0	+1·50
1300–1350	9·8	−11	−0·1	−288	81	99·5	29·3	3·2	0·5	0	+1·56
1250–1300	10·2	+2	−0·3	−293	83	103	27·8	2·9	1·0	0	+3·07
1200–1250	10·1	−13	−1·0	−290	100	102	28·6	2·3	1·5	0	+3·03

1150–1200	10·2	+26	−1·4	−292	101	103	27·5	0·0	1·0	0	+2·31
1100–1150	9·6	−43	−1·0	−281	127	99	29·3	0·0	2·0	1·0	+2·03
1050–1100	9·4 (1000–1100)	—	−0·3	−283	92	97·5 (1000–1100)	—	0·0	0	0	+1·90
1000–1050		—	−0·7	−286	93		—	0·5	0	0	+1·59
950–1000	9·2 (800–1000)	—	−1·0	−283	111	96 (800–1000)	—	1·0	0·5	0	+0·85
900–950		—	+0·1	−286	93		—	0·0	0·5	0	+0·60
850–900		—	−1·1	−285	100		—	0·3	0	0	+0·71
800–850		—	−1·1	−285	112		—	—	0	0	+2·06
750–800	—	—	−0·8	−285	107	—	—	—	0·5	0	+3·39
700–750	—	—	−1·0	−286	97	—	—	—	1·5	0	+2·81
650–700	—	—	−0·4	−285	89	—	—	—	0·5	0	—
600–650	—	—	−0·2	−283	86	—	—	—	0·5	0	—
550–600	—	—	−1·1	−286	113	—	—	—	1·0	0	—
500–550	—	—	−0·7	−288	107	—	—	—	0	0	—
450–500	—	—	−0·9	−291	97	—	—	—	0	0	—
400–450	—	—	−0·5	−291	91	—	—	—	0·5	0	—
350–400	—	—	−1·0	−289	115	—	—	—	1·0	0	—
300–350	—	—	−1·0	−285	91	—	—	—	1·0	0	—
250–300	—	—	−0·7	—	—	—	—	—	0·5	0	—
200–250	—	—	−1·8	—	—	—	—	—	0·5	0	—
150–200	—	—	−1·7	—	92	—	—	—	1·0	0	—
100–150	—	—	−0·8	—	78	—	—	—	0·5	0	—
50–100	—	—	−0·5	—	70	—	—	—	0	0	—
0–50	—	—	−0·8	—	84	—	—	—	0	0	—

(Dansgaard *et al.*, 1969) and are for the ^{18}O content of precipitation, which is mainly determined by temperature, with lower ^{18}O content associated with lower temperature. These data show a warmer interval from AD 300 to 1150 with cooler temperatures from 1200 to 1850, a minimum around 1500–1750 and a rising trend from 1850 to the present. The temporal pattern for solar activity parallels that for ^{14}C and ^{18}O (Bray, 1966a, 1970b) and both ^{14}C and ^{18}O variation have been significantly correlated with the solar activity index. The solar activity index combines sunspot and auroral observations (Bray, 1967) and, by implication with its inverse relationship to glacial advance, has a positive correlation with temperature (Bray, 1965), given a suitable lag period (Bray, 1968). The solar activity index indicates slightly lower than normal temperatures in the first 2 centuries AD and a plateau of generally warmer temperatures to the end of the 12th century with a brief and minor drop in the 7th century. Beginning in the 13th century, the solar activity data indicate there were generally lower temperatures until the mid 19th century with possible brief reversion to higher temperatures in the late 14th, mid 15th and mid to late 18th century, and then a rise in temperature to the present, especially after 1915.

3. Temperature—Human navigation and migration

Commenting on the magnitude of historical temperature changes, Lamb (1965) noted that they were small enough to account for earlier impressions that there had been no significant climatic change in the past 2500 years; but he showed that changes were still large enough to have upset human navigation, economy and agriculture, and to have caused the climate of central England to range in summer between values for the extreme south of England and the Scottish border and in winter between values for Dublin, Ireland and the Dutch coast. The effect of climate on navigation and settlement in the North Atlantic is summarized in Table IV. The warmth of the AD 900–1200 interval and the cooling which culminated in the 17th century are clearly seen from these data. As with the temperature data, the greatest changes in settlement pattern occurred at the higher latitudes: Ellesmere was settled by Eskimos in the 9th century and abandoned by 1500; Southern Greenland was settled by Europeans in the 10th century who were replaced by Eskimos by the 16th century, and Iceland was settled by Europeans in the 9th century and not abandoned, though its evacuation (Lamb, 1963) was considered. The navigation data show a similar pattern with a temperature minimum neatly outlined by the period 1605–1720 in which no European ship got through to Greenland. Following this period, navigation was resumed and resettlement by Europeans completed by AD 1894.

TABLE IV

Human settlement and navigation, North Atlantic

Date AD	Event
865	First human settlement in Iceland is made by Europeans[1]
ca. 900	First human settlement in Ellesmere Land is made by Eskimos[2]
930	Settlement of Iceland completed[1]
985	Settlement of S. Greenland by Europeans[3]
1126	Greenland gets its own bishop[3]
1203	Ice encountered on old sailing route to east Greenland near the Arctic Circle[4]
1350	Old sailing route to Greenland abandoned, new route chosen further south[4]
1377	Last bishop to live in Greenland[3]
by 1400	Height of Europeans in Greenland <150 cm,[5] soil frozen soon after interment, thus preserving clothing[6]
1410	Regular communication to Greenland stopped[4]
1476	No further travel to east coast Greenland[4]
by 1500	Ellesmere Land and N. Greenland abandoned by Eskimos[7]
1585	No Europeans found in Greenland by J. Davis[3]
1605–1720	No European ship got through ice to Greenland[4]
1695	Sea ice drifted around entire coast of Iceland[8]
1721	European resettlement on west coast Greenland[3]
1822	European ships renew contact with east coast Greenland[4]
1850	Climatic deterioration stopped contact with east coast Greenland[4]
1894	European resettlement on east coast Greenland[4]

[1] Thorarinsson, 1956.
[2] Lamb, 1965.
[3] *Encyclopedia Brittanica*, 1950.
[4] Lamb, 1967.
[5] Lamb, 1966.
[6] Ahlmann, 1953.
[7] Chard, 1962.
[8] Manley, 1961.

A source of evidence from the southern hemisphere for a possible warmer interval around AD 900–1200 is that the major period of Maori (Polynesian) emigration to New Zealand occurred around AD 900–1300 (W. Parker, personal communication, 1970). This migration represented a tropical oceanic people moving poleward into a warm temperate to temperate maritime climate, a movement which may have been greatly facilitated by the greater warmth of the period, as noted by Raeside (1948). It is an intriguing coincidence that the major Polynesian migration to New Zealand occurred at the same time as the Viking settlement of Iceland and southern Greenland; in both cases, a seafaring people may have benefited in their migration by calmer seas and in their settlement by warmer weather. The Raratongan navigator Ui-te-Rangiora is believed by tradition (Raeside, 1948) to have made voyages into Antarctic waters around AD 650, which would indicate a climate more suitable for navigation than at present.

4. Temperature—Summary

By combining information from the data in the three preceding sections with historical accounts, the following summary was constructed. Except for the substantially reduced temperatures which began after 1200, reached a minimum in the late 1600s, and then increased to the present, there have been fairly uniform temperatures over the past 2 millennia. For the first 3 or 4 centuries AD, temperatures were probably similar to the 1930s with a trough around 0–150 and a minor peak from 150 to 250. There was then a period of general stability with a slight depression around 600–700 which culminated in a period of greater warmth than the present around 900–1200, often referred to as the Little Climatic Optimum (hereafter abbreviated LCO). Temperatures then started a fluctuating decline which culminated around 1650–1700 with notable lows in the 1430s, 1560s, 1590s and especially the 1690s and early 1700s. During this decline there were reversions to milder temperatures, particularly in the late 1300s, the mid to late 1400s, the early to mid-1500s (especially) and briefly in the 1630s and mid 1600s. After the first decade of the 18th century, there was a discernible general upward temperature trend which has continued to the present with notable periods of warmth in the 1730s, around 1780, the 1830s, the 1870s and the 1930s (especially). There were reversions to cooler temperatures in the 1740s, around 1790, early 1800s (especially) around 1850, the 1890s and around 1910.

5. Precipitation

The precipitation data in Table III show that for England and Wales, there was a positive, non-significant relationship ($\rho = +0{\cdot}38$, $P > 0{\cdot}05$) between temperature and precipitation with a decline from 102 to 103% of yearly precipitation in the warmer phase from AD 1150–1300 to 92·5–96·5% in the cooler phase from 1550 to 1750. This relationship reflects the correlation of +0·42 found between annual rainfall and annual mean temperature since 1740 (Lamb, 1965). Summer precipitation however, showed an inverse trend to temperature ($\rho = -0{\cdot}27$, $P > 0{\cdot}05$), with the highest summer precipitation occurring during the 1550–1750 temperature minimum. This pattern also applies to the mean high summer wetness/dryness index for Britain, Germany and Russia near 50°N (Table III) which is correlated ($\rho = -0{\cdot}26$, $P > 0{\cdot}05$) with the winter mildness/severity index for the same area. There seems to be a tendency in these areas for colder periods to be drier on a yearly basis, but wetter during the summers, and for warmer periods to be wetter as a whole, but drier during the summers.

The positive correlation in England between total yearly precipitation and prevailing annual and winter temperatures was explained by Lamb *et al.* (1966) as being related to the prevalence of south-westerly winds. Mild epochs had a high prevalence of south-westerly winds and therefore high sea temperatures. This resulted in more moisture input into the atmosphere and moister winds blowing across the country. Colder epochs were associated with blocking anticyclones from Greenland or northern Europe and therefore less rain-produced upward motion and drier winds. Lamb *et al.* considered these relationships were probably characteristic of the western half of Eurasia in the latitudes where the westerly winds prevail and that similar conditions might be expected in Alaska and those parts of North and South America west of the Cordillera. It is significant that these areas together with the South Island of New Zealand were the ones in which the glacial advances of the LIA were very pronounced.

6. *Glaciation*

The number of weeks of drift ice off the coast of Iceland (Koch, 1945) and the number of recorded maximum glacial advances for the northern and southern hemispheres (from data compiled in Bray, 1968) are shown in Table III. These two independent estimates of glacial activity have a correlation coefficient of $+0{\cdot}87$ ($P < 0{\cdot}001$) and are very similar to the global glaciation pattern of Porter and Denton (1967) shown in Table II. These data show that almost all glaciation over the past 2 millennia occurred from 1600 to 1900 with a peak from 1700 to 1850. This period, the "Little Ice Age" was the most active of the three Neoglacial ice phases (Porter and Denton, 1967; Bray, 1970a). There was a tendency for glaciation over the past 2 millennia to be concentrated near the time of the lowest temperatures as judged by the physical and geophysical indexes, although maximum glaciation clearly lagged behind the temperature minimum. All of the physical and geophysical temperature indexes reached their minima around 1500–1700 with the 1650–1700 period the lowest 50-year interval on record, whereas the majority of glacial advances occurred from 1700 to 1900 with a maximum from 1800 to 1850 when all of the temperature indexes showed a rising trend. Indicative of this relationship is the number of years in which the Thames was frozen at London (Table II) which relates to the temperature curve by reaching a maximum during the temperature minimum in the 17th century (the last two centuries of data are not included because a bridge was built and altered the conditions of freezing).

III. Altitudinal Change in Vegetational Distribution

Change in the position of timberline, defined as the upper border of straight-boled trees, has long been assumed to be a response to climatic change. The relationship between temperature, and especially summer temperature, and timberline is the basis of this assumption (Daubenmire, 1954; Wardle, 1968), though it is apparent that many other factors may be involved. These include wind regime (Griggs, 1938; Kriuchkov, 1958), moisture (Cooper, 1942; Kriuchkov, 1958) relief forming processes (Kriuchkov, 1958) and the length of the period of snow cover which Brink (1959) found resulted in timberlines in coastal British Columbia, Canada, ranging from 1370 m near the sea to 2135 m on the drier interior slopes. Of these additional factors, however, relief forming processes change on a much longer time scale and are a constant; wind regime is related to topography and is effectively a constant over a long time period; and snow cover and moisture are correlated with temperature. This suggests that given no topographic change, an increase in temperature will result in a rise in timberline in those areas where wind or some other factor is not severely limiting.

In analysing changes in timberline it is necessary to assess the possibility of disturbance, particularly of burning or grazing. Griggs (1938) stated that "nearly everywhere" he had examined advancing timberlines in Idaho, Wyoming and Montana, U.S.A., there was evidence of old fires and that in some regions the fires were so old that "evidence of them was not to be detected without some search". Since charcoal can survive for centuries without decomposition, it is apparent that the age of the charcoal is more important than its presence, which must be widespread in most soils of the earth's surface. If this age is greater than a reasonable time for tree reinvasion it would indicate that factors other than fire have stopped regeneration in the interim.

The effect of grazing at higher elevations is substantial, whether from introduced mammals or from native mammals living with reduced or eliminated predator control. Griggs (1938) stated that an alpine meadow can be permanently changed by an amount of grazing "that would be quickly repaired on a lowland pasture" and noted that even pack horses could modify the character of the meadows. Overgrazing cannot, however, remove living or relict timberlines, but only inhibit regeneration. The worldwide upward movement of trees in the past 3 or 4 decades is clear evidence that in many areas, grazing is insufficient to stop tree migration. It was my impression of the higher elevations of the National Parks of the Canadian Rockies in the early 1960s that while there were some areas of notable overgrazing, mainly by mountain goats, these were of local occurrence and the higher elevations, as a whole, had widespread tree regeneration.

A. EURASIA

1. U.S.S.R.

In the Khibin Mountains of Murmansk district, Kozubov and Shaydurov (1965) found that at 450–520 m above sea level there were greatly weathered *Pinus* boles which reached 50–60 cm diameter at the neck of the root and were from 3 to 4 m in length. These boles were carbon dated at 600 ± 90 years. On this basis they concluded that the last maximum rise of the upper forest boundary was about 800–900 years ago. Some time after 600 BP a climatic decline resulted in the forest boundary retreating to around 270–300 m. Then, around 200–250 years ago, there was a climatic improvement with a slow upward movement of the forest boundary which has accelerated in the last 30–40 years. The present forest boundary is at 360–365 m and contains *Pinus* up to 7 m in height. Isolated individuals now occur up to 650–720 m where their seeds were probably carried by birds. Four other Russian workers are listed by Kozubov and Shaydurov as having found remains of higher timberlines.

2. Europe

A decline in timberline in the Riesengebirge of around 100–200 m from the 14th to 17th centuries was postulated by Firbas and Losert (1949) on the basis of pollen analyses. They doubted that human disturbance was responsible and suggested the decline in timberline reflected a drop in summer temperature of around 0·7–1·4°C using a July lapse rate of 0·7°C/100 m. Further west in the Swiss Alps, there was a lowering of the upper tree limit after AD 1300 by 70 m (Gams, 1937). In the present century and especially since 1930, timberlines throughout Europe have risen (Kleiselsierg, 1947; Ahlmann, 1953; Hustich, 1958). Hustich in Finland found upward expansion of *Pinus* seedlings of from 15 to 60 m (mean = 44) between 1933–1936 and 1958 and quoted papers by Erkamo and by Blüthgen who believed that the present upward tree migration was nearing the maximum post-glacial advance.

B. NORTH AMERICA

1. Canada

Griggs (1938) stated that around Lake Louise, Banff National Park, there were many large dead trees of *Larix lyallii* at the forest margin "where only small cripples are now alive" and that this recession had occurred elsewhere in the area but did not extend to Jasper National Park to the north or to Glacier Park, U.S.A., to the south. In the Canadian Rockies, Heusser (1956) searched the timberline zone for evidence of trees at formerly higher altitudes. He visited 7 locations and at 4 of these found rotting fallen tree-trunks larger in size than any nearby living trees. In 3 of these locations, fire could not be entirely eliminated as having once destroyed the tree cover. In the fourth location, east of Bow Lake in Banff Park, fallen trees were found which had branches but showed no evidence of charring. Fires had occurred on the lower slopes but no sign of fire was found on or about the fallen trees, one of which was over 60 cm in diameter near the base of the trunk. No data were given on the altitude of these trees relative to present timberline, but Bray and Struik working in the same area in 1961 observed two instances of relict timberlines of dead trees which were around 60 m above the present timberline. We did not realize the significance of these trees at the time and did not record their exact location. It is possible that they were in the locations visited by Heusser, judging by his itinerary. Brink and Farstad (1949) considered the upward advance of timberline in the Coast Range of British Columbia to reflect a milder climate since the 1750s. Brink (1959) later attributed this advance to a diminishing period of snow cover and suggested that since there were long days in July and August, an increased length of the snow-free season would be doubly significant for plant growth. Brink noted that fires had been rare in the Garibaldi Lake area and that few old trees carried fire scars. Only in one pit out of "dozens" dug was charcoal found and Brink doubted that the original Indian inhabitants would have burned the area.

Rampton (1969) working in the Snag-Klutlan area of south-western Yukon found from carbon dating that around 1220 BP, a *Picea* forest was killed by deposition of up to 1·5 m of volcanic lapilli. This forest was up to 60 m above the modern treeline. Rampton concluded that this showed that climate during the period prior to the deposition was more favourable for propagation and survival of *Picea* than at present. He noted that the presence of *Picea* seedlings or saplings above the modern treeline was not frequent enough to suggest that there had been much reaction to the recent warming trend, which was shown by the measurement of tree rings. Rampton considered that, using a lapse rate of 0·7°C/100 m, the minimum increase in mean daily summer temperature should be 0·4°C for a rise in timberline of 60 m and noted that a calculated rise of 1·1°C between 1728 and 1966 had apparently had no effect on the timberline. On this basis he suggested that the actual amount of warming required to raise the tree line may be much more than that suggested by the lapse rate.

2. U.S.A.

Griggs (1938) stated from observations of W. W. Rubey that on Middle Piney Lake south of Mt Wyoming there were numerous stumps around 60 m above the present treeline but concluded that since the area was pastured, the timberline could have been lowered as the result of grazing. On the south side of Mt Hood in Oregon, there was a dead forest of larger trees where the slopes are now bare except for straggling young trees. Griggs suggested that this could have occurred as a result of severe drought, in view of the loose soil of volcanic ash that was present. He concluded that the sporadic evidence of relict forests at higher elevations were too isolated to have climatic significance. Richmond (1962) in the La Sal Mountains of Utah found that the timberline was currently advancing upslope and encroaching over the tundra above the forest limit. Most of this advance was of *Picea* trees which were erect and uniformly developed except in exposed places and on frost-rubble deposits. A higher relict forest of dead, gnarled and twisted, relatively large trees extended further upslope from the present timberline than the upper limit of new forest growth. Richmond considered that the relict forest was killed as a result of climatic factors since there was no evidence of fire or other agents, and that it antedated the LIA. The relict timberline was at 3505 m compared with the present timberline at 3475 m.

Bamberg and Major (1968) in the Big Snowy Range, Montana, found large trunks of dead trees around 100 m (Major, personal communication, 1969) higher in the alpine zone than the present timberline. In the Uinta Mountains (110°29′W, 40°58′N), Major noted on a gently sloping alpine area that there were dead trunks of *Picea engelmannii* around 30 m above presently advancing young trees and about 90 m above the timberline of well-grown trees. Major also noted relict timberlines at three locations at Convict Creek in the Sierra Nevada (118°52′W, 37°33′N) but said that fires could have been responsible for all three occurrences. On another location near Convict Creek there was a relict timberline 100 m above presently growing trees but the topography was avalanche-inducing and this could not be ruled out as a factor. R. C. Bright (personal communication, 1969) has seen relict forests above the present timberline in the Wind River Range of Wyoming, the Beaverhead Mountains of Idaho and the Uinta Mountains and did not notice evidence of fire; he was not, however, looking for it. Patten (1963) found in the Madison Range, Montana that *Pseudotsuga menziesii* 350–500 years old were now surrounded by *Pinus contorta–Picea engelmannii* forest 100–150 years in age. This indicates that *Pseudotsuga*, which is the dominant at lower, warmer elevations may at one time have occupied the area now covered with the *Pinus–Picea* forest which suggests that the climate until AD 1450–1600 was warmer than after that time.

C. NEW ZEALAND

Burnett (1926 in Raeside, 1948) found logs of *Podocarpus totara* at an elevation of 1220 m on the Kirkliston Range, South Island. The present timberline in the area is around 1065 m, which indicates a lowering of around 155 m. On the basis of the decay rate of the less durable *Podocarpus spicatus*, Raeside estimated the *P. totara* would take 400 or more years to decay. This age combined with the life span of *P. totara* of up to 600 years indicates that the age of the logs could have been from up to 400–1000 years. This estimate is substantiated by the ^{14}C data (Molloy *et al.*, 1963) on 14 *P. totara* and *P. hallii* logs which showed an age range from 590 to 1450 years with a mean age of 770. The timberline on the Kirkliston Range may have declined, therefore, some time between AD 500 and 1350, and most likely around 1200. On the basis of this and other evidence, Raeside concluded that at some time in the past the timberline was at least 150 m higher than at present and that this difference represented a temperature change of 1·5–2·0°C.

Charcoals of *Nothofagus fusca* occur at 1128 m on the Crown Range, South Island which is 213 m above the present altitudinal limit of this species (Molloy *et al.*, 1963).

An example of a recent upward migration of timberline (P. Wardle, personal communication, 1966) has occurred in the South Island where 15 m above a *Nothofagus menziesii–N. solandri* var. *solandri* forest there is a young grove of these trees which are under 2 m in height and estimated to be probably less than 20–40 years old.

IV. Latitudinal Change in Vegetational Distribution

A. EURASIA

The influence of man on Arctic forest limits through cutting, burning, and grazing of reindeer, has been marked, though erratic in time and space (Hustich, 1966). Human populations over the past 4 centuries have increased in some areas and decreased in others while some areas have been little influenced by man until recently. In spite of these population fluctuations and the changes in rate of forest destruction which they entail, there was a generally stationary phase or southward movement of the forest limit in the 19th century and a strong northward advance in the 20th century, especially after 1920. This would suggest that while many areas of the Arctic are in the process of recovery from past disturbance, these areas are not extensive enough to obscure the evidence of recent forest advance.

1. British Isles

Lamb (1964) considered that the net decline in the extent of forest cover in the Beinn Eighe Nature Reserve in Scotland around 1300–1500 (Durno and McVean, 1959) may to some extent reflect the cooling following the LCO.

2. Finland

The forest limit in Finland retreated around 2500 years ago and has readvanced most recently since the 1920s–1930s (Hustich, 1948). Until the mid 1920s it was generally accepted that the forest limit was stable or retreating. Nordfors in 1924 was credited by Hustich (1948) as the first to note the northward advance of *Pinus*. Since then this advance has been extensively documented. Hustich attributes this advance to the warm summers beginning in the 1920s when conditions for regeneration were very satisfactory. He found a simultaneous increase in radial growth of *Pinus* and in crop production.

3. U.S.S.R.

A southward trend of timberline in the Soviet Arctic apparently started around 2000 years ago and resulted in a retreat of 2–4° in latitude (Tikhomirov, 1963). The map in Tikhomirov (1963) by Andreev shows a retreat of from 160 to 320 km in the area between the White Sea (Arkhangelsk) and the Urals (Anderma). This retreat was still occurring in the early part of the 20th century, but a reversal occurred about 1920 and the forest has now readvanced 200–700 m per year and in some areas to a total of more than 50 km (Uspenskii, 1963). Between the White Sea and the Urals, the readvance has been from 40 to 190 km (Tikhomirov, 1963). Tikhomirov (1962) states that Tyulina in 1936 was the first to produce convincing evidence that the forest is now advancing into the tundra by observing the active reproduction, increasing vitality and vigorous growth of *Larix dahurica* in the Anadyr and Khatanga basins.

B. NORTH AMERICA

1. Boreal

After the northward advance following drainage of glacier dammed lakes around 3500 BC, forest in the Keewatin district moved south from at least 63°N to south of 61°N around 1500 BC (Bryson *et al.*, 1965). By the LCO around AD 1000, the forest had again advanced to at least 61°30′ or 62°N. Around AD 1050 the forest was burnt and failed to regenerate and a Caribou Eskimo Culture appeared in the area. A tentative reconstruction of the northern limit of continuous forest in the Ennadai Lake region of Keewatin (Nichols, 1967) showed a northward advance of over 100 km which began around 250 BC and terminated at the LCO, around AD 950. After that there was a retreat south of over 150 km which culminated around AD 1550. This has been followed by an advance of over 30 km to the present day. That similar movements may have occurred in eastern Canada is suggested by the work of Terasmae and Anderson (1970) who found that *Pinus strobus* occurred at Val St Gilles, Quebec in 5030 ± 130 BP, which is 97 km north of its present limit.

Richardson in 1851 (from Raup, 1937) on the lower Coppermine River found isolated clumps of dead trees scattered over the "barren grounds" and concluded the forest line was retreating southward unaccompanied by any reproduction. He noted that while living trees were confined to sheltered places, the dead ones were on exposed hillsides where they could have grown only if the climate were more congenial. Johansen in 1919 and 1924 (from Raup, 1937) later confirmed Richardson's observations. Porsild (1938) noted that on Richard's Island there were well-preserved roots and stumps of a spruce forest 97 – 113 km north of the present forest limit and found *Larix* cones in the Mackenzie Delta 80 km north of the present range of *Larix*. Tyrrell (1896) observed isolated groves of timber north of the present timberline on the north-west coast of Hudson Bay. Ritchie (1960) found stumps and wind-felled trunks of spruce in tundra in the Caribou Lake Region of Manitoba and suggested that the forest had been destroyed by fire. Fernald (1911) found forest remains with stumps up to a foot or more ($>0{\cdot}3$ m) in diameter east of Blanc Sablon at the Straits of Belle Isle. He suggested that this forest was no longer living in the 16th century since Cartier made no mention of it. Hustich (1939) found forest remnants in Labrador beyond the present forest limits and in some cases ruled out fire as a possible cause. He concluded that a change in climate unfavourable to conifer forest was still proceeding. Hustich also suggested that the forests along the Labrador coast were taller, more luxuriant and probably stretched further north during the Viking period, since the Vikings (see Ingstad, 1969) made trips to Labrador for timber and it is highly improbable that they would have obtained their timber from the interior, if the conditions had been the same as at present.

The preceding data show that the general pattern of retreat and stagnation of the Arctic timberline during the 19th and early 20th centuries which occurred in Eurasia applied to North America as well. The subsequent reversal of this trend began (or was first noticed) in different areas at varying times. As recently as 1941, Raup concluded that the treeline was advancing in Alaska, retreating in north-western Mackenzie, approximately stable in south-western Mackenzie and apparently retreating in eastern Canada.

Advance of forest in the Kodiak, Alaska region was reported by Griggs (1934) who doubted if fire or human interference had been influential. In one instance, an island which was originally named Bare Island had, in 1867, a surface "rolling and varied with herbage and a few scattering patches of trees". In 1930, the island was covered with forest and the age of the oldest tree was 108 years. In the vicinity of Kodiak, Griggs studied a forest with open grown "mother" trees from 280 to 320 years old which had been invaded by a dense forest around 100 years ago. The appearance of limbs close to the ground on the mother trees supports Grigg's assertion that fire had not been an important factor in this forest migration. In 1948, Marr found that the forest on the east coast of Hudson Bay was actively spreading from its existing footholds, but doubted if climate was a factor in preventing the northward spread of trees since temperatures at present were above the minimum for tree existence. This could mean, however, that a climatic amelioration had occurred and the treeline had not yet caught up with it. Maycock and Matthews (1966) studied a *Salix* forest with a maximum age of <100 years in northern Ungava, Quebec, 480 km north of the present treeline. This forest had a marked increase in growth and had spread into the surrounding tundra in the 1930s and 1940s. On this basis, they suggested northern Quebec had, since the end of the 19th century, experienced a climatic amelioration similar to that in Labrador and other parts of Quebec.

2. *Temperate North America*

An evaluation of the relative importance of climate and of human disturbance on the distribution of temperate vegetation is badly needed. Pending this, it is increasingly apparent that disturbance, and especially fire, has been a fairly continuous factor in the post-glacial distribution of flammable and unprotected vegetation. Because of this continuity, change in vegetation distribution in North America probably more often represents climatic change than change in disturbance pattern. This is based on the supposition that regardless of whether a given society of Amerindians were primarily hunting or primarily agricultural, they probably started fires purposefully or by accident with reasonable regularity. These fires, combined with those started by lightning occurred sufficiently often to ignite most flammable unprotected vegetation when it had accumulated enough tinder and had a reasonable areal extent. It is conceivable that the effect of burning was more severe when attempts were made to stop fires or when firing was infrequent and so lengthened the period of fuel accumulation which ensured that the burn, when it inevitably came, was more destructive. Bromley (1935) believed that it could be safely assumed that fire has been a constant attendant of dry woods from time immemorial.

There have been some fairly widespread shifts of temperate vegetation since the Hypsithermal. In the Great Basin, *Quercus turbinella* occurred 418 km north of its present range during the Hypsithermal when it hybridized with *Q. gambelii* and later died out leaving its sterile F_1 hybrid to persist by vegetative proliferation (Cottam *et al.* 1959). A similar contraction of more southerly elements is shown by the intrusion of boreal forest into the deciduous hardwood zone in Manitoba (Löve, 1959) and Alberta (Moss, 1932; Bird, 1961) in the past 2000–3000 years, though Moss has indicated that in recent times, European man has shifted the balance in favour of boreal forest by preventing widespread burning. The treeline in the Cypress Hills of Alberta advanced beyond its present position between 1250 and 520 years ago, as shown by carbon dating of buried soil profiles (Jungerius, 1969). These dates are within the period of northward expansion of forest near Ennadai Lake (Nichols, 1967) and presumably represent an expansion during the warm period culminating in the LCO which was later terminated by the post LCO cooling. There has been a southward movement of major forest regions in Wisconsin of from 65 to 100 km since the Hypsithermal (Curtis, 1959). In the past "several thousand" years there was an advance of more stable forest in Wisconsin with an increasing content of species which terminated successional development. This was thought to reflect a moister (presumably cooler) climate, or decreased incidence of fire, or both.

In spite of the post-Hypsithermal decline of the grasslands and the more fire-resistant forest, like the *Quercus* savanna, the stable terminal hardwood forests have never regained their former extent, as shown by the present distribution of terminal forest herbs in grasslands throughout the Prairie Peninsula (Gleason, 1912; Bray, 1957). Further west, evidence is accumulating that during glacial times there was much more extensive forest in some areas of the Great Plains and that up to 200 BP fire or increased aridity or both had resulted in the elimination of some forests and the confinement of others to fire-protected scarps (Wells, 1970).

On Bear Paw Point, at Lake Itasca, Minnesota I observed in 1969 a northern conifer-hardwood forest on a fire-protected peninsula jutting into the lake. This forest had a herbaceous layer which was similar to that of the southern hardwood forest which lies 30 km to the west (corresponding to a N-S orientation since the prairie to boreal forest transition here has a W-E direction). On a presence basis, the understory flora was 61% southern, 16% northern and 29% boreal, as these terms are used by Curtis (1959) in neighbouring Wisconsin. On a density basis, the similarity to southern hardwood forest was more marked at 71% southern, 20% northern and 9% boreal. Some species characteristic of southern forest included *Allium tricoccum*, *Caulophyllum thalictroides*, *Menispermum canadense* and *Sanguinaria canadensis*. It is conceivable that these southern remnants have survived since the LCO or even the Hypsithermal; pollen analysis of a nearby bog is necessary. In northern Minnesota, Buell and Gordon (1945) noted that "islands" of the boreal *Picea-Abies* forest had been expanding at the expense of the more southerly *Acer-Tilia* forest. This expansion may have reflected the long period of cooling of the LIA, but whatever its origin, no new major expansion had occurred for the 40–50 years prior to 1955. Around 1940, the islands began to disintegrate (Buell, 1956). In 1969, I observed that this disintegration was continuing and that the *Picea-Abies* islands were being actively invaded by the more southerly species, *Tilia americana* and *Fraxinus pennsylvanica* which were often the dominant reproduction. Raup (1937) considered that within the past 200–300 years, some of the more southerly elements in the forests of central New England had been partially eliminated and that more northerly or transitional hardwoods were increasing.

C. AFRICA

In Uganda, Dale (1954) on the basis of forest distribution found a generally drier climate than today from AD 0 to 1200, with generally wetter conditions from 1200 to 1920 with a peak around 1400–1600. The first half of the 19th century, and perhaps 200 years before that, was somewhat droughty. Since 1920 warmer, drier conditions have prevailed.

D. NEW ZEALAND

From an analysis of terrace structure and soil pattern on the South Canterbury Plains, South Island, Raeside (1948) found the following phases: (1) After European settlement in the mid 19th century, there was a destruction of vegetative cover from fire and overgrazing that led to a new erosion cycle with aggradation which filled some channels and in places spread beyond the river bed. (2) Before settlement a period of stability occurred when the dominant plant association was tussock grassland. (3) Prior to this degradation there was a period of active sedimentation. (4) This sedimentation was preceded by a period "when the supply of waste was small, constant and fine in texture, the vegetation consisted of forest and the climate was warm and moist". Since the South Canterbury downlands were covered with the remains of forest, including logs and the dimpled surfaces caused by tree tip-up mounds, it is quite possible that the forest period occurred within the past millennium. Raeside estimated, on the basis of the decay rate of *Podocarpus totara* logs, that the change from forest to grassland occurred from around AD 1350 to 1450. Molloy *et al.* (1963) found a range of age in *Podocarpus* wood from AD 510 to 1300 for the South Canterbury area. On the basis of the Raeside estimate and the ^{14}C data of Molloy *et al.* (1963) it is conceivable that the 4th phase of forest cover and fine sedimentation was coincident in its latter part with the LCO. The period of active sedimentation in phase 3 must have occurred, therefore, during the post-LCO cooling when a fall in temperature exposed a considerable area to soil erosion as the uppermost limits of vegetation receded down the mountainsides. The period of erosional stability immediately preceding European settlement (phase 2) may then have corresponded with the warming after the 17th-century temperature minimum. Evidence from soils and climatic indexes (Raeside, 1948) suggested there had been a change from a wetter cycle to the present drier or more fluctuating climate. Thus, Alexandria, South Island, with a rainfall of 33 cm should, on the basis of present climate, have had Pedocol soils, but in fact had Pedalfers, which are typical of moister climates.

Holloway (1954), on the basis of changes in forest distribution, composition and regeneration pattern, suggested that in the South Island after AD 1200 there was a substantial temperature drop and, in the drier eastern section, an undoubted fall in effective regional rainfall. Much of his evidence came from a "regeneration gap" in gymnosperm trees which was shown by a predominance of larger and older individuals and a paucity of seedlings, saplings and young trees. Holloway's work was supported by the detailed investigation of Wardle (1963) at six localities in the South Island which showed that regeneration of *Dacrydium cupressinum*, *Podocarpus spicatus* and *Libocedrus bidwillii* declined around AD 1300, reached a minimum from around 1600 to 1800 and increased again around 1800–1900. Wardle ascribed this decline in podocarp regeneration more to a decrease in effective precipitation than to a temperature decline. A regeneration gap caused by presumed climatic cooling was found in Taranaki, North Island, around 1650 by Nicholls (1956) with the first and hardest hit area occurring above 610 m. A similar regeneration failure in West Taupo, North Island, forests owing to colder and drier conditions was noted by McKelvey (1953) to have occurred after AD 1600. A regeneration gap was also noted (Grant, 1963) for the Huiarau Range in the North Island around AD 1500 and possibly as far back as AD 1250. From 1550 to 1650 conditions on mid and lower slopes were again favourable for forest regrowth. A climatic interpretation of these changes was not attempted. Around 1700 there was a decrease in precipitation effectiveness which was considered to have been accompanied by a general increase in temperature.

The great importance of catastrophic disturbance (mainly fire) as a major factor in forest regeneration was raised by Cumberland (1962) and Molloy (1969), both of whom saw little reason to assume the regeneration gap was the result of climatic change. The regeneration gap was demonstrated, however, in stable forests which, while of varying age, had not been subject to complete catastrophic destruction in some cases for periods of up to at least a millennium, as shown by the size and age of their oldest trees. If the regeneration gap was the result of surface fires destroying reproduction, it is necessary to explain why the gap culminated around AD 1650–1750, which was over 750 years after the initiation of the main wave of Polynesian migration and over 350 years after its conclusion. Cameron (1964) found that a great increase in the use of fire to destroy forest by Polynesians came after the potato had been introduced in the late 18th century. Fire was used to clear land for a system of shifting cultivation of the potato, which produced a food for home use and for trade with the Europeans. This forest destruction occurred over very large areas and came at a time when gymnosperm regeneration had increased (Wardle, 1963), which suggests that a climatic change may have been responsible for the increase in regeneration after about 1800. The discussion of Molloy (1969) emphasized the endogenous nature of forest succession and suggested that periodic rejuvenation was a prerequisite for the continued existence of podocarp forests. If such succession is solely endogenous, it is difficult to imagine why the onset and the closure of the rejuvenation gap occurred in widely different parts of the two islands at about the same time.

Evidence of former higher temperatures in New Zealand was found by G. N. Park (personal communication, 1970) at an elevation of 790 m in the Tararua Mountains, North Island, in the form of a Polynesian cooking (hangi) pit, the base of which was below the water table and gave a carbon date of AD 1227 ± 52. Since the oven could not have been used if it were under water, it was inferred that the climate at that time was warmer with greater evapotranspiration. There were identifiable *Nothofagus menzeisii* roots intertwined with the hangi material which suggested the site was drier at the time the hangi was made and that forest deterioration had not yet commenced. The site at present is so infertile that a forest could not establish itself, and there is evidence that forest deterioration owing to increased wetness and wind has been occurring for at least 150 years.

V. Tree Growth

A. Introduction

There is broad agreement that tree growth close to altitudinal or polar limits is mainly controlled by temperature (Andersen, 1955; Eklund, 1957; Giddings, 1943; Laitakari, 1920; Mikola, 1952; Ording, 1941; Oswalt, 1960; Siren, 1961; Strand, 1962). This control is exerted during the current season especially in the early part (Eklund, 1957; Giddings, 1943; Laitakari, 1920) and to a lesser extent, by temperature of the preceding season (Fritts, 1966; Ording, 1941; Siren, 1961) and preceding decade (Giddings, 1943; Siren, 1961). Eklund (1957), however, found that temperature during the preceding year had a weak negative effect on growth. Temperature during the warmest part of the day is apparently of greatest importance (Huber, 1948; Strand, 1962). No study has given any indication of a clear influence of precipitation on growth, though it is evident that drainage is important in site quality (Andersen, 1955).

At lower elevations and latitudes, the control of temperature declines and precipitation is of increasing importance in regulating tree growth (Huber and Jazewitsch, 1956; Mikola, 1952; Slåstad, 1953). Thus, Slåstad found that growth at the bottom of a valley in Norway was influenced by rainfall, while on the sides and up to timberline, temperature was the main factor. In most temperate areas, growth is undoubtedly related to a number of climatic parameters (Shulman and Bryson, 1965) of which temperature appears to be more important in the early part of the growing season (Friesner, 1943; Fritts, 1958) and precipitation, soil moisture and evaporation stress more important in the mid to late growing season (Fritts, 1958, 1966; Shulman and Bryson, 1965). Where soil moisture is replenished by late summer or early autumnal rains, and temperature is still adequate, autumnal growth reaches a seasonal maximum or near maximum (Fritts, 1966).

In semi-arid sites, precipitation is the predominant factor in tree growth, though temperature may modify its influence and multiple correlations using both precipitation and temperature usually show a marked increase in predictive ability (Fritts, 1965a). As a general conclusion, Fritts (1965a) suggested that for semi-arid regions, the wider the ring, the more moist and cool was the climate, though temperature appears to influence growth only if moisture is present in the soil.

B. RELATIVE TREE GROWTH INDEXES

The diversity of methods used to correct tree ring measurements for age trend give results which need careful interpretation in studies of past climates, especially on an interregional basis. There are two basic correction techniques: those in which growth over the whole period is presented relative to growth over a preceding period (Rampton, 1969) or a limited portion of the total period (Fritts, 1965a), and those in which growth is presented relative to a regression curve or an "eye-fitted age trend curve" (Haugen, 1967).

An example of growth relative to a previous period of uniform length is that of Rampton (1969) who divided each annual ring width by the mean ring width of the previous 10 years. Fritts (1965a) chose a "standard" interval of from AD 1651 to 1920 and all growth was related to this interval. The advantage of these techniques is their ease of calculation. A greater amount of calculation is needed to fit a regression curve and divide each ring width by each yearly value of the fitted curve. The advantage of such fitting is that all ring widths are transformed to indexes with a mean of 1·00 and have a variance which is independent of tree age, position within the stem and mean growth of the tree (Fritts, 1966). Another advantage is that indexes receive equal weight when averaged.

The disadvantage of relative indexes for long-term climatological study is that they can be used to show a relative increase or decrease in growth only over a fairly short-term period. This means that if the climate is less favourable for tree growth for a period of several consecutive centuries and then more favourable for several more centuries, the tree ring indexes will show only the variation within these centuries, not the great contrast between the two periods. Tree ring indexes thus measure a hazy area somewhere between weather and climate and their use in climatic reconstruction has resulted in conflicting and generally inconclusive results. If growth is presented relative to a limited period in which there was a uniform climate, then it may be possible to assess the effect of climatic change. This was, fortunately, somewhat the case with the study of Fritts (1965a), who used the period 1651–1920 as a basis for growth comparison, a period that was contemporaneous with a generally cooler than average portion of the past millennium. The study of Brehme (1951) also seems to have revealed tree growth change over the longer-term period by the use of wood from buildings of a preceding age, thus reducing the effect of age trend from the use of only living trees.

C. ABSOLUTE TREE GROWTH INDEXES

The use of absolute tree growth indexes was attempted by Bray and Struik (1963) who measured growth on an ecosystem basis by sampling every tree in the stand and calculating the total stand bole growth. The calculation was made with the simplifying assumption that each tree bole was composed of a series of perfect concentric circle segments, the area of which was calculated by measurement of total radius and the width of each radius segment. The result was expressed in cm^2/year of bole increment. Measurements of decadal intervals of the increment cores were made in the field to avoid the distortion in growth pattern which may occur with storage of cores and the resultant differential drying and shrinkage. By sampling each tree in the stand it was assumed that variation in individual tree growth owing to competition or relative age would be eliminated and only total stand growth portrayed.

D. TEMPERATURE TRENDS

Brehme (1951) analysed living *Larix europea* at high elevations near Berchtesgaden and by correlating these with wood from cowherds' huts was able to prepare a chronology from AD 1340 to the present. His study was made to determine if the decline of treeline proposed by Firbas and Losert (1949) from the 14th to 17th centuries could be verified by tree growth patterns. The results showed a steep decline in growth after 1430 with a minimum from 1580 to 1740, a strong recovery from 1750 to 1780 and a lesser recovery in the mid to late 19th century. Brehme considered these results supported Firbas and Losert's since the growth increment was reduced from AD 1400 to 1600 by nearly half the relative value. A study with a somewhat similar aim was made by Fürst (1963) who analysed *Quercus* growth in the centuries preceding and following 1650 which he considered was the beginning of the LIA. Fürst calculated an index based on the ratio between two growth statistics, which he used as an indicator of cold winters. The index was 0·61 for AD 1280–1380, 0·67 for 1550–1650, 0·76 for 1650–1750 and 0·65 for 1850–1950, thus demonstrating greater cold during 1650–1750 than previously.

Bray and Struik (1963) in a study of *Picea engelmannii* growth in Yoho National Park, British Columbia, Canada found a positive χ^2 relationship ($P < 0{\cdot}005$) with the world temperature data of Willett

(1951) from 1740 to 1960 and a negative χ^2 relationship ($P < 0{\cdot}01$) with precipitation from 1820 to 1920. It was concluded that the negative growth–precipitation relationship was probably the result of the strong negative correlation between Willett's temperature and precipitation regimes. There was also a significant negative relationship ($P < 0{\cdot}001$) between forest growth and the number of glacial advances in north-western North America from 1580 to 1960. Since the glacial data were closely correlated with the Willett world climate regimes it was concluded that climatic patterns in western North America may have been closely synchronous with world trends. These correlations also strengthened the significance of the relationship between forest growth and the Willett climatic data.

Adamenko (1963) plotted the weighted tree ring index of Schove (1954) for Scandinavia against the growth of *Larix sibirica* in the polar Urals and concluded the results were strikingly similar if there was a 25-year lag between the data. He tentatively concluded that this change in phase was due to a large-scale change in atmospheric circulation and that delay in the growth of Scandinavian trees was explained by the delay in the frequency of anticyclonic situations in western compared with eastern Europe. Adamenko found more favourable conditions for tree growth in the 18th century and in the 20th century after the 1920s. These patterns were also noted by Bray (1966b), who found that Canadian data were similar to those of the polar Urals with higher growth in the mid to late 18th century, and following the first two decades of the 20th century. A further circumpolar comparison by Haugen (1967) found highly significant correlations for 19 of 23 50-year intervals between 1650 and the present for his own data on *Picea glauca* from interior Alaska and for the polar Urals data of Adamenko, the Schove Scandinavian index and the Hustich data from Labrador. He considered the basis of the similarity in tree growth was summer (June or July) temperatures.

Tree-growth indexes for 9 stands in sub-arctic or sub-alpine forest are summarized by 50-year intervals in Table V. The results are somewhat inconclusive. In general, growth was moderately high in the 20th century (especially after 1920 or 1930), lower in the 19th century, particularly the early part, notably high in the 18th century especially from 1750 to 1800, generally low in the 17th century, quite divergent in the 16th and 15th centuries but notably high in the 14th century. That these trends are broadly related to the long-term temperature trends in Tables II and III is apparent. Of particular interest is the similarity in tree growth between 1750–1800 and 1930–1960 for the studies of Bray and Struik (9·7 *vs* 9·6), Haugen (119 *vs* 115) and Adamenko (19·0 *vs* 19·5) which closely support the ^{18}O and solar activity

TABLE V

Tree growth indexes: sub-arctic and sub-alpine forest

Reference	Brehme 1951	Siren 1961	Eidem* 1953	Eidem* 1953	Ording* 1941	Erlandsson* 1936	Adamenko 1963	Bray & Struik 1963	Haugen 1967
Region	Germany	Fenno-scandia	Norway	Norway	Norway	Sweden	Polar Urals	Canada	Alaska
Species	*Larix europea*	*Pinus silvestris*	*Picea abies*	*Pinus silvestris*	*Pinus silvestris*	*Pinus silvestris*	*Larix sibirica*	*Picea engelmannii*	*Picea glauca*
Data	r**	r	r	r	r	r	r	cm^2/year	r
1900–1950	—	+30	—	—	—	—	14·9	9·3	110
1850–1900	1·27	−40	+0·42	+2·24	−1·40	−0·64	13·9	9·2	102
1800–1850	0·94	−210	+1·02	+0·58	+1·70	−1·56	16·0	8·6	92
1750–1800	1·48	+80	+2·00	+0·68	+2·04	−0·08	19·6	9·7	119
1700–1750	0·82	+225	+0·42	+1·02	−1·68	−1·10	13·7	8·9	137
1650–1700	0·69	+50	+0·82	+2·26	+0·34	+0·30	—	8·7	120
1600–1650	0·70	−160	−1·10	−1·40	−2·14	−1·98	—	—	—
1550–1600	0·90	+110	+1·70	+0·80	+1·36	−0·54	—	—	—
1500–1550	0·78	+50	+2·26	−0·04	−2·70	−1·32	—	—	—
1450–1500	1·41	−90	+1·07	−0·87	−0·55	−1·40	—	—	—
1400–1450	1·89	−70	—	—	—	—	—	—	—
1350–1400	1·96	+30	—	—	—	—	—	—	—
1300–1350	—	+180	—	—	—	—	—	—	—
1250–1300	—	+140	—	—	—	—	—	—	—
1200–1250	—	−100	—	—	—	—	—	—	—
1150–1200	—	−100	—	—	—	—	—	—	—

* Data from Schove (1954). ** Relative.

data in Table III and suggest that the warmth of the late 18th century was similar to that of the mid 20th.

A summary of the data in Table V is shown on a mean rank basis in Table VI and demonstrates the growth patterns outlined above and in particular emphasizes the high growth of the 1750–1800 period. These conclusions support those of Adamenko (1963), Bray (1966b) and Haugen (1967) on the general similarity of circumpolar sub-arctic forest growth.

TABLE VI

Tree growth indexes: rank order

	Sub-arctic and sub-alpine forest[1]						Arid forests[2]		
No. studies	4	3	9	8	6	2	2[a]	3[b]	5[b]
1900–1950	2·5	3·5	—	—	—	—	—	—	—
1850–1900	4	5	2	4	4	6	4	3	1
1800–1850	5	6	4	5	7	10	6·5	5·5	5
1750–1800	1	1·5	1	1	1	1	5	8	3
1700–1750	2·5	1·5	3	3	5	4	8	8	4
1650–1700	—	3·5	—	2	3	9	10	10	6
1600–1650	—	—	—	—	9	11	11	11	7
1550–1600	—	—	—	—	2	3	3	1	2
1500–1550	—	—	—	—	8	8	1	2	—
1450–1500	—	—	—	—	6	7	9	5·5	—
1400–1450	—	—	—	—	—	5	6·5	8	—
1350–1400	—	—	—	—	—	2	2	4	—

[1] Data from Table V.
[2] Data from Fritts (1965a).
[a] Arizona only
[b] Arizona + New Mexico.

Tree growth in semi-arid stands can also be used to portray temperature trends if the conclusion of Fritts (1965a) that "the wider the ring, the more moist and cool was the climate" is correct. The inverse of the semi-arid stand chronologies was used, therefore, as a temperature index with low tree growth representing higher than average temperatures. Since Fritts' data cover a wide geographic, altitudinal and climatic range, an analysis was made (Table V) of similarity in Fritts' tree-ring chronologies for different regions and elevations in comparison with the sub-arctic and sub-alpine forest data. It was found that the closest rank correlation existed between the most arid of the lower elevation stands of Fritts and the tree-ring chronologies in Table V. For the longest period for which data are available from at least 2 sub-alpine or sub-arctic stands in Table V, AD 1350–1900, there was a rank correlation of $+0{\cdot}55$ ($P < 0{\cdot}05$) between the mean of the Brehme and Siren rank data and the mean of the inverse of stands 9 and 10 of Fritts, both from Arizona in south-western U.S.A. A high correlation also existed between the Brehme and Siren data and stands 7 (New Mexico), 9 and 10 of Fritts, but it was not statistically significant. On the assumption that the inverse of stands 9 and 10 represents temperature trends before 1350 (for which data in sub-arctic and sub-alpine stands are lacking), a temperature chronology for these stands is presented from AD 700 to 1950 in Table III. The similarity of this tree-ring chronology to general temperature trends is moderately close, especially if the data are smoothed in a running mean.

In spite of some lack of similarity in the tree-ring chronologies in Table V, there are close similarities in the patterns of alternating higher and lower values in the original data, if these patterns are considered independently of absolute maximum and minimum values. Summaries were made of these alternating patterns for each of the stands in Table V, and these summaries were then combined into the following pattern: there were high values around 1480–1490, low values from 1530 to 1540, high around 1560, low from 1590 to 1620, high from 1630 to 1660, low from 1650 to 1720, high from 1740 to 1825, low from 1820 to 1850, high from 1850 to 1900, low from 1910 to 1920 and high from 1940 to 1950. The reason that the numerical patterns in Table V are not more similar and that the above sequence of higher and lower values are remarkably similar is probably that the techniques used to correct for age trend render the relative tree-ring indexes less comparable but cannot obscure the sequences of greater and lesser values. This result augments the studies of Adamenko (1963), Bray (1966b) and Haugen (1967) in suggesting that the chronologies in Table V, which were drawn from both North America and Europe, may reflect generally synchronous global temperature trends.

E. PRECIPITATION TRENDS

Fritts (1965b) presented 26 tree ring chronologies for western N. America, the oldest being from AD 576. Analysis of relative 10-year departures since 1500 showed widespread drought in 1576–1590, 1626–1635, 1776–1785, 1841–1850, 1871–1880 and 1931–1940. Above average moisture occurred during 1611–1625, 1641–1650, 1741–1755, 1826–1840 and 1906–1920. Before the 16th century there were not sufficient tree samples for detailed analysis.

VI. Herbaceous and Agricultural Growth

A. EURASIA

1. Iceland

According to Thorarinsson (1956), barley and oats were cultivated from the time of first settlement in AD 865. Cultivation of cereals had ceased in the north and east by 1200 and had diminished in the south and south-west during the 13th and 14th centuries. By 1350, barley only was still grown in a few places in the south and all grain growing ceased by 1500 except for one district, where it persisted until the end of the 16th century. In the 17th, 18th and 19th centuries, grain growing experiments were made which were sometimes successful, especially in the south and south-west but were never continued. Lamb (1967) noted that the warmth of the 1730s was accompanied by a remarkable upturn in human population; after 1750 there was a climatic recession and population decline which, again, may have been related to grain production. Since 1923, oats and barley have been grown regularly in the south and since 1930 in all parts of the country.

In one of the most isolated parts of Iceland, grain growing was abandoned in the 14th century and a native shore lyme grass *Elymus arenarius* was harvested as a cereal. This plant, which yields a small crop and a poor-quality flour, would not have been grown if better flour were available and this fact is used by Thorarinsson as an indication that the decline in cereal production at that time was not due to changes in trade conditions. The destruction of agricultural fields by advancing glaciers occurred in many areas of Iceland. In 1695 a farm was abandoned which was buried by a glacier in 1708 and in 1698 another farm was abandoned which was reached by ice 4 years later (Ahlmann, 1953).

In the 20th century, Iceland has shared in the general warming trend with consequent changes in plant and animal distribution (Thorarinsson, 1956). Peat and ice hummocks, "rusts", which are the same type as the palses of Lapland and which are found in the marshes of the interior, are gradually disappearing as a result of the milder climate. There has also been a rise in the lower marsh limit which corresponds to the northward recession of the southern limit of the Siberian permafrost zone (Ahlmann, 1953).

2. *Greenland*

Evidence of early Norse burials in south-western Greenland, and plant roots deep in ground now permanently frozen, suggest that mean annual temperatures at the peak of the LCO must have been 2–4°C above present (Lamb, 1965). By the 14th and 15th centuries, the permafrost was close to the surface (Vebaek, 1962), which presumably means tillage was no longer possible. The decline in size of cowhouses from the 13th to 16th centuries is another indication of a decline in plant growth. The role of human disturbance in this decline has been summarized by Ingstad (1966) who noted that the use of wood for domestic fuel and the smelting of bog-ore together with the grazing of woodlands led to their disappearance over large areas. As a consequence, the soil dried and rain washed away the turf and topsoil, exposing the land to wind erosion. In some cases the soil was blown away and sand-drifting occurred. The loss of trees was particularly damaging because the lushest grass often grew under the trees or in the lee of them. Ingstad believes, however, that the despoiling of vegetation was not sufficiently great to have caused the disappearance of the European settlers. The recent warming trend is illustrated by the desiccation of hummocks in north-eastern Greenland during the past 75 years as shown by the ages of willows growing on the hummocks (Raup, 1966).

3. *Norway*

The limit of farming in central Norway between AD 1000 and 1320–1350 was 100–200 m higher in the uplands than at any time before AD 800 or for long afterwards (Halmsen, 1961, in Lamb, 1967). Wheat had grown as far north as Trondheim from AD 1150 to 1300 (Lamb, 1969a). Corn prices fell from about 1300, though it was not until 1348 that the Black Death first appeared (Steensberg, 1951). There was famine (Table VII) in the 1590s and 1690s in the diocese of Stavanger; in 1596–1598 the famine was so bad that people ate bark and ground hay, straw and chaff into bread (Utterström, 1955, in Lamb, 1959). According to Hoel and Werenskiold (1962), the climate was relatively good in 1667 since a new and higher tax was worked out in Jostedal which might indicate a certain degree of prosperity. From 1680–1690 to 1743, however, glaciers advanced down many Norwegian valleys and destroyed meadows, pastures, tilled fields and farm buildings. The farm Tungöen in central western Norway fed 38 cows and 3 horses and produced 29 barrels of grain in 1667 (Rekstad, 1901). In 1702, 1728 and 1735, taxes on Tungöen were decreased because of declining production. In 1728, the farmhouse was moved because of fear of an overhanging glacier, which later went over a hanging valley on 12 December 1743 and killed all people and farm animals except 2 boys and 2 cows. In 1740, 1741 and 1742 the climatic deterioration was so severe, according to Hoel and Werenskiold (1962), that barley and oats did not ripen, resulting in widespread famine. In 1740–1741 deaths exceeded births by 31 000 out of a population of 700 000.

4. *Finland*

Erkamo (1956) has compiled observations on the northward expansion of herbs following the warming trend in the 20th century. Hustich (1948) also noted a northward advance in Lapland of more delicate kinds of grain during the 1920–1939 period. Hustich (1952) found strong positive correlations between temperature and quantity of rye crops

TABLE VII

Periodicity of volcanism, poor harvests owing to excessive cold or moisture, and poor tree growth

Major volcanic eruptions[1]		Poor harvests[2]				Poor tree growth[3]
AD	Dust Veil Index	Japan	Scotland	Norway	England and Wales	(range of minima)
1902	1000	—	—	—	—	1910–20
1883	1000	—	—	—	—	—
1875–78	1550	—	—	—	—	—
		1866–69	—	—	—	—
1845–46	1250	—	—	—	—	—
1835	4000	—	—	—	—	—
		1833–39	—	—	—	1820–50
1831–33	1000	—	—	—	—	—
1815*	3000	—	—	—	—	—
1807–10	1500	—	—	—	—	—
1803	1100	—	—	—	—	—
1783–86	1150	1782–87	1780s	—	—	—
1772–75	1250	—	—	—	—	—
1763–68	4000	—	—	—	—	—
1752–55	1700	—	—	—	—	—
		—	—	1740–42	—	—
1693–94	1500	—	1690s	1690s	1693–1700	1650–1720
1680	1400	—	—	—	—	—
no data before 1680		—	1590s	1590s	1590s	1590–1620
	—	—	—	—	1550s	1530–40
		—	—	—	1430s	1440–50
		—	—	—	1314–25	no data before 1340

[1] From data in Lamb (1969b).
[2] From text in Lamb (1959, 1964), Hoel and Werenskiold (1962) and Arakawa (1955a).
[3] From all studies summarized in Table V.
* 1816 was "Year without a summer".

and concluded that the great increase in yield per hectare of rye was largely due to the temperature increases in the 1920s and 1930s. Hustich (1948) noted that relative cereal growth of rye and oats closely paralleled that for radial growth of pine. The great increase for the 1920–1939 period was quite marked, as well as the minimum values for 1900–1909, a period of generally world-wide temperature decline.

5. *Denmark*

The first record of farm abandonment in Denmark was 1334 (Steensberg, 1951). For the next 300–400 years this became a common happening. From the middle of the 13th until the late 14th century in Zealand, the parish tithes of 79 parishes declined by about 70%. Most Danish villages with the termination "thorp" were founded from the 10th to 12th centuries. Nearly all villages abandoned during the late Middle Ages had the ending "thorp", which suggested to Steensberg that they were founded during the LCO on more marginal farm lands and were then abandoned when the climate worsened. In one small district of Jutland, 16 out of 34 "thorp" villages were abandoned during the late Middle Ages. In 1406, English visitors to a royal wedding reported much uncultivated sodden land and that wheat was grown nowhere (Lamb, 1967).

6. *Scotland*

In 1290, cereal grains grown and distributed by the abbey at Kelso in south Scotland consisted of 84% oats, 11% barley and 5% wheat (Lamb, 1964). In the 1590s, the first of three severe famines (Table VII) occurred in Scotland as the result of poor summers and harsh winters (Lamb, 1959). From 1693 to 1700 every kind of grain harvest failed in all but the lowest and most favoured eastern parishes in 7 out of 8 years. There were desperate attempts made to save bedraggled, unripe corn in brighter, drier weather in January, but in spite of these there was widespread famine. Many parishes were reduced to a half or a third of their inhabitants (Lamb, 1959, 1964). In 1733, the growing of wheat was reintroduced in the lowlands of west Fife around Dunfermline, thus beginning the recovery period though a subsequent climatic reversion in the 1780s also led to famine (Lamb, 1964).

7. *England and Wales*

The growing season was drier and warmer during the Roman era in southern Britain and the latter part of this era was one of prosperous estates, some of which survived in Somerset until well after the Roman withdrawal around AD 400 (Lamb, 1966). Following this period there was apparently a fair degree of climatic and presumably agricultural stability with evidence of cooling in the 8th century (Lamb, 1964) and of drought in the late 8th century (Schove and Lowther, 1957). During the subsequent LCO, from 1150 to 1300, vineyards in England reached their greatest extent (Lamb, 1967). These vineyards were on average 4° north of the modern limit of viticulture in France, a distance which corresponds to a difference of about 1°C in average summer temperature (Lamb, 1967). Many of the vineyards were in areas that in modern times are exposed to northerly winds and frequent night frosts in May. This led Lamb to suggest that frosts at that time were rare as the result of a warmer sea and less ice on the Arctic Ocean. One vineyard at Tewkesbury, Gloucester, was in an obvious frost pocket, again suggesting that late frosts were not a serious concern at the time. Lamb (1966) examined the present climates of the best wine districts by report in 1150–1300 and compared them with the present climate of the northern limit of commercial vineyards in Europe. He found that prevailing temperatures in the former English vineyards were 0–1·5°C below the mean lowest value in May, 0–2·5°C below extreme lowest May value, 0·2–1·3°C below mean July temperature and the same as mean October temperature. From this he concluded it was cool springs, May frosts and, to a lesser degree, lack of summer warmth that resulted in the demise of English commercial vineyards. This demise began around 1320 and continued into the late 1400s. The vineyard at Ely continued until 1469 when the accounts suggest little wine had been produced for 150 years; the annual values before 1320 were sometimes ten times as high as any mentioned after that. In the late 1400s there were scattered attempts to re-establish vineyards. These attempts have been described for one vineyard, which was re-established on a sheltered south-facing bank with high red brick walls around it, as a "rich man's fantasy" (Lamb, 1967).

The 12th and 13th centuries were a time of upward agricultural expansion with complaints in 1234 that tillage was spreading so far up the hills in the north as to leave too little land higher up for pasturage (Beresford, 1951). The highest medieval cultivations were further into marginal land than the over-enthusiastic ploughing up campaign during the food shortage years of 1940–1944 (Beresford, 1951). Remains of a 13th-century corn-drying oven were found on Dartmoor at 350 m and the 13th-century limit of tillage in Northumberland was 300–490 m which is 120–150 m above the limit of worthwhile cultivation at the present. The height difference of 150 m would correspond to 1°C warmer summers (Lamb, 1967). Beginning in the early 1300s, there is evidence of recurrent periods of poor harvests, accompanied by widespread social distress and, sometimes, famine. During these periods (Table VII), nearly every summer was short, cold, moist and overcast with usually early frosts and sometimes deep snow in autumn (Lamb, 1964, 1967). The 1590 crop failures were responsible for the Elizabethan Poor Laws. In 50 deserted village sites in Oxfordshire and 34 in Northamptonshire, there was a population decline of one-third from 1280–1300 to 1327 and the decline was 67% for those places that were later wiped out by the Black Death in 1350. Lamb (1967) suggests that the people in these places may have been undernourished as a result of poor harvests and thus more susceptible to the Black Death. The remaining sites had more people in 1377 than before the Black Death and reached their final decline from 1440 to 1480. The phenomenon of deserted villages in the late Middle Ages has been frequently interpreted as the result of increased land use for sheep pasturage with consequent eviction of the original farmers. Lamb (1967) outlines numerous examples of decline in rural populations and abandonment of towns beginning in the early 1300s to the mid 1400s and continuing until the late 1600s; it is unlikely that all this rural decline was due to the expansion of sheep farming.

A period of partial climatic recovery occurred during 1400–1550 (Lamb, 1959) which, except for the crop failures in the 1430s, saw somewhat better harvests and the expansion into English gardens of southern fruits like apricot, fig, peach and quince; cherries appeared about 1460. Following the disastrous famines of the 1690s, gradually improving agricultural conditions in England and Wales have reflected the general temperature increase since that time, though the 1740s, 1780s, 1810s and 1820s were also a time of climatic and presumably agricultural recession. Turner (1964) noted that small-holdings in Cardiganshire reached a height of 300–335 m during the mid to late 18th century, a period of notable warmth (as seen in Table III). The later abandonment of many of these holdings was ascribed by Turner to the agricultural depression following the Napoleonic wars. It is conceivable that some of this abandonment was the result of the low temperatures of the 1810s and 1820s as shown by Manley (1961).

8. Germany

There was a general lowering of the upper limit of cultivation of fruit and grain crops after 1300 (Gams, 1937). Wine production, which had occurred in East Prussia as early as 1128, was stopped in 1437 (Lamb, 1959). In Baden, the upper limit of vineyards was lowered by 220 m after 1300–1430 which implies a mean temperature before the lowering that was 1·5–1·6°C higher than around 1930 (Müller, 1953, in Lamb, 1965). Wine harvests in Baden from 850 to 1950 (Table II) illustrate the general cooling from the LCO to minimum periods in the 17th and 19th centuries, with recovery in the 16th, 18th and 20th centuries.

9. France

There was a permanent retraction in the limits of grape cultivation in France as a result of cooling from around 1550 to 1700. The harsh winter frosts of 1709 killed the vines over wide areas and cultivation in many places was never resumed. Between 1550 and 1800 there was a retardation of 20 days in the average date of grape picking in the Vivarais district (45·7°N, 4·8°E) near Lyons (Lamb, 1965, 1967).

10. Switzerland

Extensive glacial advances occurred in Switzerland during the LIA; those in the 17th century overwhelmed mountain villages, several of which are still covered by ice (Porter and Denton, 1967). Agricultural production was stopped or greatly decreased by these advances though it is interesting that W. Coxe in 1776 (Onthank, 1951) found that Mer de Glace glacier of Mont Blanc had extended "far into fields of corn and rich pasture" and lay "without being melted, in a situation where the sun had sufficient power to ripen the fruits of the field: it is literally true that with one hand we could touch ice, and with the other ripe corn".

11. U.S.S.R.

The former presence of a more northerly forest has been extensively documented by the use of relict herb and shrub species (Tikhomirov, 1963). These forest elements did not disappear with the post-Hypsithermal cooling but became a part of the tundra and survived in sites with favourable microclimates. They remained strictly confined within the former forest limits. In the past 40–50 years, a northward movement of up to 0·5° latitude has occurred which has accompanied the general northward forest migration.

12. China

The richly recorded history of China undoubtedly contains much material on plant growth. I only located the notation by Lamb (1967) that repeated frost damage caused the abandonment of orange and tangerine growing in the Kiangsi Province (25–30°N) around 1654–1676.

13. Japan

In the summer of 1180, there was a bad drought in western Japan which at that time was ruled by the Taira family. By analysing the diaries of three noblemen, Arakawa (1960) found that there had been 14 rainy days in June, none in July and 4 days with showers in August; data were unknown for 7 days in July. These figures were compared with the droughts of 1883, 1924 and 1939 at Kyoto during which there were an average number of rainy days of 13 in June, 10 in July and 8 in August. There was so little rain in 1180 during the normal rainy season from around 10 June to 10 July that it was impossible to plant rice. The resulting widespread death from starvation led to the conquest of the Taira regime by invaders from eastern Japan where there had been a bumper rice harvest.

During the Tokugawa era (1603–1868), there were three notable periods of poor harvests, 1782–1787, 1833–1839 and 1868–1869 (Table VII), which resulted from cold summers with excessive rain, occasional hoar frosts and, usually, a predominant north to north-east wind (Arakawa, 1955a). The years 1783, 1786, 1833, 1836, 1866 and 1869 had particularly bad harvests with famine often resulting. A compilation of rice yield data from Tochigi (north of Tokyo and directly south of Tohoku District) and the price of rice in Kyoto was computed from data in Arakawa (1955a) relative to the decadal means (Table VIII). The periods of nationwide famine are reflected in these data and, in particular, the famine of 1836 which had a yield of 26% of the decadal mean and was followed by a year in which the price of rice was 158% of the decadal mean. The worst famine area was usually the northern district of Tohoku, one part of which had a death rate of 9% and an emigration rate of 4% from October 1783 to May 1784. Rice in northern Japan requires a July and August mean temperature of not less than 20°C and the 9 bad harvests in Tohoku since 1883 were all caused by cold wet summers. Between 1902 and 1940 in Tohoku, the correlation coefficients between rice yield and mean air temperature were $+0{\cdot}85$ ($P < 0{\cdot}001$) for July and $+0{\cdot}79$ ($P < 0{\cdot}001$) for August.

There are a series of dates (Arakawa, 1955b) for the cherry blossom festival at Kyoto which extend from AD 812 to 1864. These dates are difficult to interpret since their distribution is clumped and three different species of cherry are involved. The festival was held at the time of full bloom and could have varied for social or political reasons relative to the time of the first appearance of full bloom (which lasts for these species for longer than a week in New Zealand at a comparable latitude). If these dates are averaged by 50-year intervals then for all intervals with 5 or more dates there is only 8 days difference between the extremes and no indications of a temporal trend.

B. NORTH AMERICA

During the LCO, there was a widespread occurrence of predominantly agricultural people (Mound Builders) south of the Great Lakes and into southern New England. These people entered the more northerly state of Minnesota around AD 1000 (E. Johnson, personal communication, 1969) bringing with them beans and squash. They were gradually replaced from the Middle-West from the 12th to 14th centuries (Aschmann, 1962) and in New England from 1350 to 1450 (Raup, 1937) by people more dependent on hunting. Raup considered that a warmer and drier climate in southern New England at the time of the Mound Builders would have favoured agricultural development since com-

TABLE VIII

Yield (y) and price (p) of rice relative to decadal mean in Tochigi (y) and Kyoto (p), Japan (calculated from Arakawa, 1955a)

	0	1	2	3	4	5	6	7	8	9
1770*p*	—	—	—	0·96	0·87	0·93	1·06	1·13	1·05	1·00
1780*p*	0·61	0·72	0·94	1·12	1·16	0·89	0·97	1·68	1·00	0·90
1790*p*	0·85	0·87	1·12	1·11	0·90	0·99	1·07	1·03	1·03	1·01
1800*p*	1·22	1·12	1·05	0·98	0·80	0·81	0·88	0·93	1·10	1·11
1810*p*	0·91	0·95	0·91	0·93	1·11	1·13	1·04	1·15	0·99	0·87
y	0·98	0·90	1·02	0·93	0·90	0·91	0·97	0·96	1·23	1·22
1820*p*	0·73	0·83	0·95	0·95	1·02	1·03	1·14	0·94	0·99	1·35
y	1·07	0·96	1·06	1·07	1·04	0·75	1·11	0·93	0·99	1·04
1830*p*	0·82	0·87	0·78	0·98	1·09	0·78	1·07	1·58	1·06	0·98
y	1·12	1·19	0·99	0·66	1·21	1·16	0·26	1·18	0·93	1·30
1840*p*	0·82	0·86	0·96	0·88	0·98	1·05	1·10	1·06	1·09	1·19
y	1·08	0·96	1·01	0·76	1·12	0·85	1·11	0·97	1·11	—
1850*p*	1·14	1·15	0·85	0·95	0·99	0·82	0·79	0·92	1·19	1·19
y	0·82	0·92	0·98	1·13	1·04	0·99	1·01	1·00	1·04	1·08
1860*p*	—	—	—	0·33	0·42	0·75	1·74	1·67	0·85	1·22
y	1·09	1·07	0·93	1·15	1·06	1·19	0·85	0·96	0·93	0·81
1870*p*	1·26	0·74	—	—	—	—	—	—	—	—
y	0·96	1·12	1·01	1·06	0·99	0·91	1·08	0·95	1·06	0·86
1880*y*	1·02	0·81	0·98	0·93	0·91	0·93	1·19	1·08	1·09	1·06
1890*y*	1·00	1·05	0·91	1·04	1·16	0·95	0·90	—	1·04	1·02
1900*y*	0·99	1·20	0·78	1·07	1·13	0·78	0·85	1·04	1·06	1·09
1910*y*	0·96	0·97	0·91	1·01	0·92	1·16	0·96	0·97	0·98	1·16
1920*y*	1·19	0·93	0·91	0·88	0·98	1·03	0·98	1·14	1·05	0·92
1930*y*	1·11	0·92	0·98	1·11	0·92	1·04	1·04	1·10	0·72	1·02

petition between agriculture and forest reinvasion for the occupancy of cultivated lands would have been less rigorous than later under a climatic cooling. It is likely that the northward expansion of predominantly agricultural people up to about AD 1000 was related to the warmth of the LCO and the subsequent retreat from the 12th to 15th centuries was the result of the climatic cooling at that time. Perhaps a major factor in this retreat was the increasing occurrence of cold summers with crop reduction or failure.

In the Great Plains there was an apparent contraction of agricultural people from the west to east from the 11th to 16th centuries (Aschmann, 1962). In the south-west of the U.S.A. there was an expansion of agriculture during the 11th century with a subsequent contraction from the late 13th to 16th centuries. In desert conditions in south-western New Mexico and south central Arizona, there was a retraction from 600 000 to 200 000 km^2 although this shrinkage of area was only accompanied by an estimated population decrease of one-tenth. These conditions were assumed to be the result of a decrease in moisture during the growing season. Any interpretation of North American population levels is difficult because of the great changes which took place with the invasion of Europeans.

Among native plant populations there has been a degeneration of *Carex* sod in the high elevation of the central Sierra Nevada, which was assumed by Klikoff (1965) to indicate that in terms of the last several hundred years, a drier period was entered about 1915 and this has continued to the present. Bamberg and Major (1968) found remnants of turf 30–35 cm thick occurring in rock rubble in the Big Snowy range in Montana. They noted that flat, bare rock outcrops had solution channels which probably developed under a turf but offered no explanation of their origin.

C. New Zealand

Raeside (1948) found that close tussock vegetation ceased at around 1829 m in the Two Thumb Range of the South Island. From 1829 to 1981 m there were residual patches of the original tussock which were no longer in equilibrium with their environment and were being driven back by encroaching scree. The area was lightly stocked with sheep and there were infrequent fires; disturbance was, therefore, presumed not to be a major factor. Soil types similar to those occurring from 1676 to 1829 m were found between 1829 and 1981 m mixed with rocky debris. These observations were used to suggest that plant cover once extended upward at least 150 m.

VII. Discussion

A. Synchroneity

Changes in plant growth and vegetational distribution for which a climatic explanation is possible appear to have been broadly synchronous throughout the world over the past 2 millennia. A similar global climatic synchronism was demonstrated on a statistical basis for glaciation in North America, Eurasia and the Southern Hemisphere over the same interval (Bray, 1968). On a longer time-scale, climatic synchronism between the Northern and Southern hemispheres was shown for the Wisconsin cold period by Epstein *et al.* (1970). In compiling the information for the glaciation analysis, I was continually impressed by coincidences in the timing of LIA advances and of recent retreats for greatly diverse parts of the world; similar coincidences have occurred in analysing the vegetational distribution and growth data. For example, the pattern of cooling, following the LCO, is shown by all of the vegetational distribution and plant growth data, though the European tree altitude distribution changes were evident at an earlier stage than those in North America. The North American herbaceous (agricultural) data, however, are similar to the European data and indicate that cooling began from the 12th to 15th centuries. In the Southern hemisphere, some studies indicate cooling as early as the 12th century, while others find no evidence until the 15th century. Considering that the general cooling, shown by the physical data to have begun in the 12th–13th centuries, was interrupted by warming trends, it is not surprising that the influence of this cooling on plants was evident at different times throughout the cooling phase, depending on specific variability and local climates and, in part, on the accuracy of the botanical chronologic techniques. A lag in climatic change might occur between Eurasia and the southern Pacific area. Emiliani (1970) suggested that the difference in temperature between glacial and interglacial was 5–6°C for equatorial Atlantic and 3–4°C for equatorial Pacific, an inequality which is related to the larger volume and greater thermal inertia of the Pacific Ocean. Ericson and Wollin (1970) have concluded that some periods of warm surface waters in the Pacific were coincident with cool water in the Atlantic.

The botanical data also show a global synchroneity in their response to the warming trend which has occurred since the temperature minimum in the late 17th century. Change in Arctic tree line, while synchronous for Eurasia and North America, is mainly evident since the 1920s, perhaps because observations in the Arctic were not often recorded until the 19th century, a period of generally lower temperature and forest retreat, thus missing the possibility of observations on a possible northward movement during the warmth of the late 18th century. In Nichols' (1967) study, however, the southward retreat of boreal forest in Keewatin was found to have halted around AD 1550 and an advance of over 30 km followed at sometime thereafter.

B. Relation to Climate

Temperature trends indicated by the botanical data parallel those of the physical indexes with a cooling beginning in the 13th century, a very minor recovery evidenced by a few studies in the 15th or 16th centuries, a minimum in the 17th century and a subsequent increase to the 20th century with a secondary minimum in the 19th. The European studies show that a decline in timberline began around AD 1300–1400 with a subsequent reversal and upward movement within the past 200–250 years and most rapidly since 1930. In North America, the only dated decline was 1220 BP and it occurred following volcanic destruction from which, however, there was no subsequent recovery. In other studies, the condition of preservation of the relict timberline suggested it had existed before the LIA (Richmond, 1962; Heusser, 1956; Bray and Struik, 1963) as does the date of 1450–1600 (Patten, 1963) for the replacement of *Pseudotsuga* by *Pinus–Picea.*

On the assumption that the undated North American relict timberlines were pre-LIA and probably reached their maximum around the LCO, the lowering of timberline between AD 900–1300 and 1550–1700 was 30–100 m in North America, 70–200 m in Europe, and possibly 150 m in New Zealand. Given a lapse rate of 0·7°C/100 m, this represents a temperature decline of around 0·2°C to 1·4°C which is less than the maximum temperature difference of 1·5°C in Table III and of 1·9°C in Table II. This lower estimate of temperature decline based on timberline change is in accord with Rampton's observation that the amount of temperature change needed to change the treeline is greater than that suggested by the lapse rate.

The LIA decline in timberline of from 30 to 200 m can be compared with a decline of around 50–500 m (Gavelin, 1909; Samuelsson, 1910; Hustich, 1948; Pears, 1968; LaMarche and Mooney, 1967) from the Climatic Optimum to the present. These data support the observation that climatic change over the past millennium has been nearly as great as since the Thermal Maximum.

Assessment of the relationship of tree growth to climate is hampered by the paucity of data before AD 1350 and by difficulty of correcting for age trend. The inverted and smoothed curve (Table III) for stands 9 and 10 of Fritts (1965a) reflects the higher temperature of AD 700–850, 1050–1350 and 1500–1550 and the generally lower temperatures from 1550 to 1950. Similarly, the combined Brehme–Siren data (column 6, Table VI) show higher growth from 1350 to 1400 and from 1750 to 1800 with the lowest growth from 1600 to 1700 and from 1800 to 1850. Before 1250, when only Siren data are available, the tree growth data suggest lower temperatures which is not in accord with the physical data. The conclusions of Brehme (1951) and Fürst (1963) that there was lower tree growth after 1600–1650 than before are evidence for the influence of post LCO cooling. The period of increasing temperature after the post LCO minimum in the 17th century is reflected by all the tree growth data in Table V which show a growth increase to maximum or near maximum values in the late 18th century, a recession in the 19th century and, in most cases, a recovery in the 20th century (especially after 1930).

The information on herbaceous and agricultural growth consistently reflects the temperature trends of the past millennium. The history of grain cultivation in Iceland, for example, showed an optimum growth from the time of settlement in AD 865 to around 1200 followed by a decline to 1550 and virtual cessation until 1730, with partial recovery after that time and complete recovery since 1930. This pattern is repeated by those from Scandinavia and parts of the British Isles. Continental European growth patterns also show the general temperature trends, though I was not able to find as much information on this area as for the North Atlantic countries. The Japanese data do not extend for a sufficient length to determine whether there was a post LCO cooling trend, while the one piece of information from China suggests cooling in the 17th century. Given the difficulty of reconstructing historical agricultural growth in North America, it is impressive that the data from the Middle West and New England both seem to indicate a decline in warmth following the LCO. How much the interpretation of the evidence for this decline has been influenced by knowledge of European trends is not possible to assess, though work being done at present on Amerindian population levels by mass excavation of entire villages may give a firmer foundation to climate–growth patterns for this period. The information showing warming and drying in native vegetation at higher elevations in the 20th century is in accord with the temperature patterns.

The data on poor harvests owing to excessive cold or moisture (Table VII) encompass every period of notably cold weather which was recorded during the post-LCO cooling except the "year without a summer", 1816. The lack of famine after 1700 in England and Wales, after 1742 in Norway and after 1780 in Scotland may be the result of better political organization, food storage and distribution rather than a climatic phenomenon, just as the famine of 1866–1869 in Japan had the partly political origin of the breakdown of the Tokugawa feudal era. The compilation of alternating periods of relative lower and higher tree growth in Section V is shown in Table VII to give patterns which are generally synchronous with the poor harvest data.

A compilation has been made of all information from the botanical data in the present review which indicate either relative warmth or a warming trend, or which indicate relative coolness or a cooling trend. The results (Table II) clearly show the same general pattern as the physical indexes, although the data before around AD 800 are meagre and of usually lesser reliability. The compilation in Table II gives only a relative indication of climate but may, nevertheless, emphasize some aspects of climatic change more fully than shown by the physical indexes. One of these is the sharp contrast between the warmth of the 11th and 12th centuries and the cooling of the 13th and especially 14th centuries. Another is the unanimity of the assessment of the coolness of the 17th century during which not one piece of botanical information indicates either relative warmth or a warming trend. The extent of the climatic recovery of the 18th century is shown by an increase to 47% of the botanical information implying relative warmth. The depth of the recession of the 19th century is highlighted by a decline to 16% of values implying relative warmth. The warmth of the latter half of the 18th century is emphasized by all of the high latitude and high altitude tree growth data which indicate a higher temperature at this time than during the first half of the 20th century, an assessment which is in agreement with the ^{18}O data.

The botanical data strongly support the physical indexes in showing that the timing of the temperature decline following the LCO was not exactly coincident with the LIA. Rather, the physical and botanical data (see especially columns 1, 3, 4 Table III) indicate that cooling began in the 13th century, reached a minimum in the late 17th century and was followed by a general warming trend to the present. Lamb (1965) stated that this cold epoch culminated from 1550 to 1700. The LIA is not really discernible (column 2 Table II, columns 8–10 Table III) until 1550–1600; it reached its maximum around 1700–1850 and it was still of a fair magnitude in the early 20th century. Since the post-LCO cold epoch is often designated as the LIA or at any rate identified with it in a temporal sense, it might be helpful if this period were labelled with a separate name (Little Climatic Minimum?).

The preceding analyses show that there is a certain sensitivity of vegetation to climatic change on the order of centuries and millennia. This sensitivity is in marked contrast to the usual failure to establish vegetation–climate relationships based on yearly or decadal means which has led some scientists to consider climatic change of little significance in interpreting vegetation dynamics. As an awareness grows of the possibilities of using botanical data to assess climatic changes of the recent past it is certain that present techniques will be more widely and accurately applied and that other paleobotanical methods will be invented. When this occurs, it is likely that a more subtle picture will be available of the period of temperature decline which followed the LCO and of the nature of the recovery which has taken place in the past few centuries and especially since the 1920s.

C. BOTANICAL DATA AND CLIMATIC MECHANISMS

The physical, geophysical, glaciologic and botanical data summarized in Tables II and III show generally similar temporal patterns which are synchronous on a global scale for the cooling which followed the LCO and the warming after the LIA. This synchroneity adds evidence to the possibility that there is a physical control over climate which operates on a global scale. The most likely form for such control would be variation in solar radiation retention or receipt by the earth or variation in solar output.

1. Solar radiation retention

No atmospheric mechanism is known from the pre-industrial age which could have changed the amount of radiation retained by the atmosphere, but there is the possibility that changes in earth surface albedo could have affected retention at ground level. Kukla (1969) has suggested that an increase in the mean winter albedo of Europe, which occurred as a result of large-scale deforestation, may have been the cause of cooling from about AD 1200. He also suggested that a similar result may have occurred from the large deforestation in the U.S.A. in the 18th and 19th centuries. This suggestion is not supported by the history of European forest clearance which does not show a sufficient increase in the rate of clearing during the LCO to have resulted in a global climate change which gave the lowest temperatures since the sub-Atlantic minimum and perhaps since the Cochrane–Cockburn readvance. Change in albedo in the U.S.A. after European settlement had a complex pattern with a distinct possibility that winter albedo is lower now than before AD 1600. An increasing amount of ecological research has shown that owing to repeated natural or man-set fires, much of the pre-European forest area of the U.S.A. was in a degraded condition and would therefore have had a higher winter albedo than closed canopy forest. In many areas, the amount of land under closed canopy forest is greater now than before European settlement. In the grassland–forest border area, a study of a transect from an agricultural area to the centre of a large city in Minnesota, U.S.A. (Bray *et al.*, 1966) showed that while summer visible albedo had increased 15%, there was a possibility that winter visible albedo was only slightly higher than before European settlement. In an investigation into winter snow conditions in the grassland portion of Minnesota, I found that winter albedo was probably lower than before European settlement, as a result of late autumn and early spring cultivation which exposed heat-absorptive dark soil surfaces to the zenith. A further factor was the production of wind-blown dust from the ploughed fields which, after a light snow had melted, blew over and darkened nearby snow-covered surfaces. It is conceivable that at present the forested areas of the mountains, coastal plains and Great Lakes of the U.S.A. together with the grasslands have a lower autumn (especially), winter and spring albedo than before European migration (and perhaps considerably lower in the case of the grasslands, many of which have converted to forested surfaces, as in the Aspen parklands). The increasing areal density of man-made structures like cities and, especially, highways which have a lower winter albedo than vegetated areas because of snow removal is another significant factor in winter albedo decline.

2. *Solar radiation receipt*

A major factor influencing the receipt of solar radiation at the earth's surface is volcanism. A graph of great volcanic eruptions in Iceland and of polar ice off Iceland at 20-year intervals from AD 860 to the present showed an apparent association between glaciation and volcanism (Lamb, 1969b). Another series of graphs (Lamb, 1969b) showed similarities in a 25-year cumulative volcanic Dust Veil Index (DVI) and various world, European and English temperature series. These similarities are supported by data in the present review for the period AD 1680 to the present for which the DVI is available. The data on poor harvests in Table VII show that 3 of the 5 poor harvests since 1680 followed within one year of major volcanic eruption and one of these, the famine around 1782–1787, was directly attributed to volcanism by Arakawa (1955a). Of the remaining two, the meagreness of the 1866 and 1869 harvests in Japan was complicated by the internal troubles at the end of the Tokugawa feudal period. In the famine of 1740–1742, however, barley and oats did not ripen in Norway for three successive years, an unlikely random event but one easily explained by a preceding volcanic eruption with a heavy dust veil lasting for three years. The extensive ice advances in the 1740s and early 1750s and the 1740–1742 famine both suggest that a major volcanic eruption occurred around 1739 which has not yet been historically verified.

Lowered temperatures following volcanism were originally demonstrated by Abbot (1913) and Humphreys (1920). Temperature departures from the world temperature data of Köppen for 1750–1871 and Humphreys for 1872–1920 presented in Arakawa (1955a) are shown in Table IX for the 2 years preceding and the 5 years following the 13 major volcanic eruptions with a DVI of 1000 or greater. In the first

TABLE IX

World temperature departures preceding and following 13 major volcanic eruptions, AD 1750–1920
(data of Köppen and Humphreys from Arakawa, 1955a)

Year	−2	−1	0	+1	+2	+3	+4	+5
No. of positive departures	8	6	8	2	4	6	8	5
Mean temperature departure °C	+0·07	0·00	+0·08	−0·31	−0·27	−0·07	+0·15	−0·28

year after eruption, only 2 out of 13 temperature departures were positive and in the second year after eruption, only 4 were positive. A χ^2 test applied to the 2 years preceding (14 of 26 positive) and to the 2 years following eruption (6 of 26 positive) gave a value of 7·5 ($P < 0{\cdot}01$). Mean world temperature in the 2 years following major volcanic eruption was 0·3°C lower than average.

The number of glacial advances was compiled from data of Bray (1968) for the 10 years surrounding the start of each 4-year interval in which there was a DVI of 1000 or greater. The 10-year period was used because of the time lag between ice accumulation and maximum glacial advance (Bray and Struik, 1963) and a minimum estimate of ± 5 years in the accuracy of many of the glacial advance dates. Results of χ^2 analysis showed there were 64 glacial advances in the 133 years surrounding the first year of the 15 volcanic eruption intervals compared with 41 advances in the 148 remaining years ($\chi^2 = 7{\cdot}8$, $P < 0{\cdot}01$). This correlation of glacial advance with volcanism is presumably the result of the colder temperatures and the well-documented decrease (Lamb, 1969b) in solar radiation at the earth's surface which occur in the years following volcanic eruption. This correlation, together with the evidence of colder and shorter summers as indicated by the data on poor harvests, suggests that a decrease in the length of the glacial ablation season may be a major factor in LIA high elevation ice accumulation and advance. It is a characteristic of shorter ablation seasons that precipitation, which normally falls at the beginning and end of the season as rain, will nearly all fall as snow (Manley, 1961). It is probable, therefore, that glacial advance is stimulated by the greatly increased albedo of icefields as a result of increased frequency of snowfalls during the ablation season (albedo new snow > 80%; albedo old dirty ice < 20%). This conclusion is supported by Sauberer and Dirmhirn (1950) who made a monthly study of radiation balance at 3000 m in the Alps and found that reflection from the surface was the most variable factor and that frequent light snowfalls caused a negative radiation balance.

The influence of volcanism on climate seems to be mainly a short-term one with volcanic eruption causing temperature decline and decrease in plant growth in cooler climates and perhaps regulating the periodicity, though not the occurrence of glaciation. There is little indication that volcanism affects longer term climatic change. This is demonstrated by the cumulative DVI dates since AD 1500 in Fig. 12 of Lamb (1969b) which reached a peak around 1760–1880, a period that was generally coincident with a rising temperature trend as shown by both the geophysical and physical data in Table III, though the early 19th century was probably not much warmer than the late 17th.

Furthermore, the initiation of the period of cooling and vegetation change in the 13th century following the LCO was apparently not primarily the result of volcanism, judging by the Icelandic volcanism data of Lamb (1969b). Other evidence that volcanism does not create the conditions for glaciation is the recurrence of post-Pleistocene glacial phases at regular intervals of around 2600 years (Bray, 1968, 1970a) which may be a cyclic phenomenon for which there is no indication that the more erratic volcanic periodicity was responsible.

Receipt of solar radiation at the top of the atmosphere is influenced by astronomical change in the earth's orbit (Milankovitch hypothesis). This hypothesis cannot be used to explain the cooling and subsequent warming of the past millennium since temperature change over this period according to the Milankovitch hypothesis should be uniformly downward (Kukla, 1969).

3. Solar radiation output

Variation in solar radiation output may have been a major climatic influence over the past 2 millennia, judging by the work of Lawrence (1950), Willett (1951), Bray and Struik (1963) and Bray (1965, 1968) on the relationship of solar activity to glaciation. An analysis of glaciation in Alberta and British Columbia, Canada, and Washington and Oregon, U.S.A. (Bray and Struik, 1963) showed that from 1580 to 1900, two intervals, 1711–1724 and 1835–1849 contained over half of the glacial advances. These intervals followed the two lowest periods of solar activity (1645–1715 and 1798–1833) since 1610. Periods of high solar activity were followed by glacial stagnation or retreat. A further analysis of these data (Bray, 1965) with additional information showed that given a lag period of 18 years to allow for ice accumulation and flow to the terminus, there was a χ^2 of 7·7 ($P < 0{\cdot}006$) for the hypothesis that an equal number of glacial advances should occur following the 4 highest and 4 lowest sunspot activity periods since 1611. A summary of world-wide glaciation over the past 2000 years compared with an index of solar activity (Bray, 1968) found that glaciation was associated with periods of lower solar activity and deglaciation with periods of higher solar activity providing these periods were of sufficient length and were not preceded by long intervals of greatly differing solar intensity.

The botanical data reflect the general pattern of solar activity variation over the past 2 millennia (Table III). The summary of botanical trends in Table II shows: (1) a generally warmer interval during the higher solar activity of the 1st millennium AD which culminated around 1200–1300 and was followed by (2) a cooling trend correspondent with the general lower solar activity from the 13th–19th centuries and (3) a warming trend to the present correspondent with the general return to normal solar activity.

A comparison was made on a century basis between the quantitative geophysical, physical, glaciologic and botanical indexes in Tables II and III and the solar activity index using Spearman's rank correlation coefficient and the *t* test for large samples (Siegel, 1956, p. 212). Since there is an apparent lag effect in the influence of solar activity on climate, perhaps due to the thermal inertia of the oceans (Bray, 1968), a comparison was also made with a cumulative solar activity index which included data for each preceding century. The results, shown in Table X, indicate that 3 of the 12 geophysical, physical, glaciologic and botanical indexes were significantly correlated with the solar activity index and 11 out of 12 were significantly correlated with the cumulative solar index. These results extend the previous analyses of Bray and Struik (1963) and Bray (1965, 1967, 1968) and provide a quantitative basis for a close relationship between solar activity and a wide range of climatic indexes. The increase in the statistical significance of the solar-climate correlations, when a lag of 100 years was incorporated in a cumulative solar index, suggests that if oceanic thermal inertia delays climatic response to solar activity change, then a cumulative period of less than 200 years is necessary for the oceans to readjust to change in solar heat input. When solar activity changes are of short duration, however, and are surrounded by long intervals of differing solar activity (Bray, 1968) there will apparently be little change in general temperature levels. This is shown by the scarcity of data indicating a strong cooling trend during the brief periods of lower solar intensity of AD 317–348, 381–426, 592–707, 881–957 and 1011–1083 and by the scarcity of data indicating a strong warming trend during the brief higher solar intensity intervals of 1359–1387, 1525–1586 and 1724–1798, although the last two intervals and especially the last one had some evidence indicating warming, perhaps because we are closer to these periods historically.

The preceding analysis supports the concept of a close correlation between solar activity and climate over the past 2 millennia. In an indirect way it also adds evidence for the relative validity of the solar activity index (Bray, 1967), a validity that was supported by correlation with the geophysical indexes of ^{18}O (Bray, 1970b) and ^{14}C (Bray, 1967). Of the various physical mechanisms which have so far been proposed for a solar-climate relationship, the theory emphasizing the importance of solar flares appears most pertinent to the results of the present review. Schuurmans (1965) found for Western Europe that there was a positive relationship between the prevalence of south-westerlies and solar flare activity, and it is notable that Lamb (1967)

TABLE X

Rank correlation probabilities of various climatic indexes with solar activity

Climatic Index	Solar index	Solar index cumulative 200 years	Period centuries AD	Source of data Table
Temperature 40–90°N	<0·10	<0·0005	1–20	II
Temperature central England	>0·10	<0·025	9–20	III
Winter mildness/severity	>0·10	<0·05	12–20	III
Precipitation, England and Wales	>0·10	<0·025	9–20	III
High summer wetness/dryness	>0·10	<0·025	12–20	III
^{14}C	<0·10	<0·05	1–20	III
^{18}O	>0·10	<0·05	4–20	III
World glaciation	>0·10	<0·05	1–20	III
Ice off Iceland	<0·05	<0·025	9–20	III
Good wine harvests	<0·05	<0·0005	9–20	II
Tree growth index	<0·10	>0·10	8–20	III
Summary, botanical data	<0·025	<0·005	8–20	II

found a great dominance of south-westerly wind around 1350, 1530, 1730 and 1920, 4 dates which are at the beginning of 4 of the 5 periods of higher solar activity outlined by Bray (1967) over the same period. A decreasing proportion of south-westerlies was considered by Lamb to be characteristic of drier and more open winters which are harsher than normal. Schuurmans (1969) in a later study concluded that an increase in solar flares resulted in the development of warm and dry summers in western Europe and probably prevented winters from becoming severe. These studies support the work of Yamamoto (1961) and Willett (1961), who found an increase in zonal westerlies in the high latitudes, an increase in annual air temperature and an intensification of the monsoonal type of general circulation accompanied an increase in solar activity.

Whatever the nature of the physical control of terrestrial climate, the results of the present review should provide some stimulus for further geophysical study of solar-climate mechanisms and of the effects of volcanism. Given the general thermal equilibrium of the earth over the past several millennia, solar variation, volcanism and other factors as well have probably influenced the small but notable climatic fluctuations the earth has experienced. The generally lower variability of temperatures during the first millennium may have been the result of a stable, active sun and decreased volcanism; on the other hand, the extensive cooling after the 13th century was the result of lowered solar activity which resulted in a period of decreased temperature. At the close of this period there was a major phase of glaciation in which glacial periodicity was regulated by the effects of an active period of volcanism. An interval of generally increased solar activity following the solar dearth of the late 17th century has continued to the present, accompanied by a fluctuating rise in temperature. The early mid-20th century in particular was characterized by high solar activity and almost no volcanism and was probably similar in temperature to much of the warmer periods of the first millennium and LCO.

4. *Future research*

Research areas from which more data are needed have been indicated throughout this review. The 1st millennium AD is lacking in information and more studies from this period will hopefully become available. At present our best data are probably the three geophysical indexes in Table III all of which indicate the 1st millennium was of about the same temperature as the LCO. As the number and variety of studies increase, it will be possible to take a more critical view of the data and their relationship to the physical and geophysical climatic indexes (which should also increase in number and accuracy). This should, then, enable a clearer assessment of the greatest difficulty in interpreting paleoecologic information: the relative importance of climate and of human and natural disturbance in changes in plant density and distribution.

Acknowledgements

I am grateful to Drs E. Bakusis, D. B. Lawrence, G. J. Struik and H. E. Wright, Jr., for assistance with the widely scattered literature on climate and plant response and for stimulating discussion.

References

Abbot, C. G. (1913). *Nat. Geographic Mag.* **24,** 181–198. Do volcanic explosions affect climate?

Adam, D. P. (1969). Interim Res. Rept. No. 15, Dept. of Geochronology, Univ. Arizona, Tucson. Ice ages and the thermal equilibrium of the earth.

Adamenko, V. N. (1963). *J. Glaciol.* **4,** 449–451. On the similarity in the growth of trees in northern Scandinavia and in the polar Ural Mountains.

Ahlmann, H. W. (1953). Amer. Geog. Soc. Bowman Mem. Lectures Ser 3. pp. 1–51. Glacier variations and climatic fluctuations.

Andersen, H. E. (1955). Alaska Forest Res. Center, Station Paper No. 3. Climate in south-east Alaska in relation to forest growth.

Arakawa, H. (1955a). *Pap. Met. Geophys., Tokyo* **6**, 101–115. Meteorological conditions of the great famines in the last half of the Tokugawa period, Japan.

Arakawa, H. (1955b). *Geofis. pura appl.* **30**, 147–150. Twelve centuries of blooming dates of the cherry blossoms at the city of Kyoto and its own vicinity.

Arakawa, H. (1960). *Weather* **15**(5). The weather and great historical events in Japan.

Aschmann, H. H. (1962). In Proc. Conf. Climates 11th and 16th Centuries, Aspen, Colo., June 16–24. Natl. Center Atmospheric Res. Tech. Notes 63, 18. Summary report of the anthropology section.

Bamberg, S. A. and Major, J. (1968). *Ecol. Monogr.* **38**, 127–167. Ecology of the vegetation and soils associated with calcareous parent materials in three alpine regions of Montana.

Bassett, I. J. and Terasmae, J. (1962). *Can. J. Bot.* **40**, 141–150. Ragweed, *Ambrosia* species, in Canada and their history in postglacial time.

Beresford, M. W. (1951). *Geogrl J.* **117**, 9–149. The lost villages of medieval England.

Bird, R. D. (1961). Ecology of the aspen parkland of Western Canada. Canada Dept. Agr., Publ. 1066.

Bray, J. R. (1957). *Am. Midl. Nat.* **58**, 434–440. Climax forest herbs in prairie.

Bray, J. R. (1965). *Nature* **205**, 440–443. Forest growth and glacier chronology in north-west North America in relation to solar activity.

Bray, J. R. (1966a). *Nature* **209**, 1065–1067. Atmospheric carbon-14 content during the past three millennia in relation to temperature and solar activity.

Bray, J. R. (1966b). *J. Glaciol.* **6**, 321. Similarity of tree growth in northern Scandinavia, polar Urals and the Canadian Rockies.

Bray, J. R. (1967). *Science* **156**, 640–642. Variation in atmospheric carbon-14 activity relative to a sunspot–auroral solar index.

Bray, J. R. (1968). *Nature* **220**, 672–674. Glaciation and solar activity since the fifth century B.C. and the solar cycle.

Bray, J. R. (1970a). *Nature* **228**, 353–354. Temporal patterning of post-Pleistocene glaciation.

Bray, J. R. (1970b). *Science* **168**, 571–572. Solar activity index: validity supported by oxygen isotope dating.

Bray, J. R. and Struik, G. J. (1963). *Can. J. Bot.* **41**, 1245–1271. Forest growth and glacial chronology in eastern British Columbia and their relation to recent climatic trends.

Bray, J. R., Sanger, J. E. and Archer, A. L. (1966). *Ecology* **47**, 524–532. The visible albedo of surfaces in central Minnesota.

Brehme, K. (1951). *Z. Weltforstw.* **14**, 65–80. Jahrringchronologische und-klimatologische Untersuchungen an Hochgebirgslärchen des Berchtesgadener Landes.

Brink, V. C. (1959). *Ecology* **40**, 10–16. A directional change in the subalpine forest-heath ecotone in Garibaldi Park, British Columbia.

Brink, V. C. and Farstad, L. (1949). *Canad. Field Nat.* 63, 37. Forest advance in north and central British Columbia.

Bromley, S. W. (1935). *Ecol. Monogr.* **5**, 61–89. The original forest types of southern New England.

Bryson, R. A., Irving, W. N. and Larsen, J. A. (1965). *Science* **147**, 46–48. Radiocarbon and soil evidence of former forest in the southern Canadian tundra.

Buell, M. F. (1956). *Ecology* **37**, 606. Spruce-fir, maple-basswood competition in Itasca Park, Minnesota.

Buell, M. F. and Gordon, W. E. (1945). *Am. Midl. Nat.* **34**, 433–439. Hardwood-conifer forest contact zone in Itasca Park, Minnesota.

Cameron, R. J. (1964). *N. Z. Jl For.* **9**, 98–109. Destruction of the indigenous forests for Maori agriculture during the nineteenth century.

Chard, C. S. (1962). In Proc. Conf. Climates 11th and 16th Centuries, Aspen, Colo., June 16–24. Natl. Center Atmospheric Res. Tech. Notes 63, 17. Eskimos in the American Arctic.

Cooper, W. S. (1942). *Ecol. Monogr.* **12**, 1–22. Vegetation of the Prince William Sound region, Alaska; with a brief excursion into post-Pleistocene climatic history.

Cottam, W. P., Tucker, J. M. and Drobnick, R. (1959). *Ecology* **40**, 361–377. Some clues to Great Basin postpluvial climates provided by oak distributions.

Cumberland, K. B. (1962). *In* McCaskill, M. (ed). "Land and Livelihood", Christchurch, N.Z. Geog. Soc. pp. 88–142. Climatic change or cultural interference?—New Zealand in Moa-hunter times.

Curtis, J. T. (1959). The Vegetation of Wisconsin. Madison, Wis.: Univ. Wisconsin Press.

Dale, I. R. (1954). *Empire Forestry Rev.* **33**, 23–29. Forest spread and climatic change in Uganda during the Christian Era.

Dansgaard, W., Johnsen, S. J., Möller, J. and Langway, C. C. (1969). *Science* **166**, 377–381. One thousand centuries of climatic record from Camp Century on the Greenland Ice Sheet.

Daubenmire, R. F. (1954). *Butler Univ. bot. Stud.* **11**, 119–136. Alpine timberlines in the Americas and their interpretation.

Dorf, E. (1960). *Am. Scient.* **48**, 341–364. Climatic changes of the past and present.

Durno, S. E. and McVean, D. N. (1959). *New Phytol.* **58**, 228–236. Forest history of the Beinn Eighe Nature Reserve.

Eklund, B. (1957). *Meddr Skogsforsvaes.* **47**, 1–63. Om granens årsringsvariationer inom mellersta Norrland och deras samband med klimatet.

Emiliani, C. (1970). *Science* **168**, 822–825. Pleistocene paleotemperatures.

Epstein, S., Sharp, R. P. and Gow, A. J. (1970). *Science* **168**, 1570–1572. Antarctic ice sheet: stable isotope analysis of Byrd station cores and interhemispheric climatic implications.

Ericson, D. B. and Wollin, G. (1970). *Science* **167**, 1483–1485. Pleistocene climates in the Atlantic and Pacific Oceans: a comparison based on deep-sea sediments.

Erkamo, V. (1956). *Ann. Bot. Soc. Vanamo* **28** no. 3. Untersuchungen über die pflanzenbiologischen Folgeerscheinung der neuzeitlichen Klimaschwankung in Finnland.

Fernald, M. L. (1911). *Rhodora* **13**, 109–162. A botanical expedition to Newfoundland and southern Labrador.

Firbas, F. and Losert, H. (1949). *Planta* **36**, 478–506. Untersuchungen über die Entstehung der heutigen Waldstufen in der Sudeten.

Friesner, R. C. (1943). *Proc. Indiana Acad. Sci.* **52**, 36–44. Some aspects of tree growth.

Fritts, H. C. (1958). *Ecology* **39**, 705–719. An analysis of radial growth of beech in a central Ohio forest during 1954–1955.

Fritts, H. C. (1965a). *Mon. Weath. Rev.* **93**, 421–443. Tree-ring evidence for climatic changes in western North America.

Fritts, H. C. (1965b). *Bull. ecol. Soc. Am.* **46**, 93. Tree-ring evidence for climatic changes in western North America from A.D. 1500 to A.D. 1940.

Fritts, H. C. (1966). *Science* **154**, 973–979. Growth-rings of trees: their correlation with climate.

Fürst, O. (1963). *Flora* **153**, 469–508. Vergleichende Untersuchungen über räumliche und zeitliche Unterschiede interannueller Jahringbreitenschwankungen und ihre klimatologische Auswertung.

Gams, H. (1937). *Z. dt. öst. Alpenver* **68**, 157–170. Aus der Geschichte der Alpenwälder.

Gavelin, A. (1909). *SkogsvFör. Tidskr.* no 7. Om trädgränsernas nedgång i de svenska fjälltrakterna.

Giddings, J. L., Jr. (1943). *Tree-Ring Bull.* **9**, 26–32. Some climatic aspects of tree growth in Alaska.

Gleason, H. A. (1912). *Bot. Gaz.* **53**, 38–49. An isolated prairie grove and its phytogeographical significance.

Grant, P. J. (1963). *Trans. R. Soc. N.Z.* (Bot.) **2**, 143–172. Forests and recent climatic history of the Huiarau Range, Urewera region, North Island.

Griggs, R. F. (1934). *Ecology* **15**, 80–96. The edge of the forest in Alaska and the reasons for its position.

Griggs, R. F. (1938). *Ecology* **19**, 548–564. Timberlines in the northern Rocky Mountains.

Haugen, R. K. (1967). *Science* **158**, 773-775. Tree ring indices: a circumpolar comparison.

Heusser, C. J. (1956). *Ecol. Monogr.* **26**, 263–302. Postglacial environments in the Canadian Rocky Mountains.

Hoel, A. and Werenskiold, W. (1962). "Glaciers and Snowfields in Norway". Norsk Polarinstitutt, Oslo Univ. Press.

Holloway, J. T. (1954). *Trans. R. Soc. N.Z.* (Bot.) **82**, 329–410. Forests and climates in the South Island of New Zealand.

Huber, B. (1948). *Naturwiss.* **35**, 151–154. Die Jahresringe der Bäume als Hilfsmittel der Klimatologie und Chronologie.

Huber, B. and Jazewitsch, W. von (1956). *Tree-Ring Bull.* **21**, 28–30. Tree-ring studies of the Forestry-Botany Institutes of Tharandt and Munich.

Humphreys, W. J. (1920). "Physics of the Air". McGraw-Hill, Philadelphia.

Hustich, I. (1939). *Acta Geogr., Helsingf.* **7**, 1–76. Notes on the coniferous forest and tree limit on the east coast of Newfoundland–Labrador.

Hustich, I. (1948). *Acta bot. fenn.* **42**, 1–75. The Scotch Pine in northernmost Finland and its dependence on the climate in the last decades.

Hustich, I. (1952). *Fennia* **75**, 97–105. Agricultural production in Finland and the recent climatic fluctuation.

Hustich, I. (1958). *Fennia* **82**, 1–25. On the recent expansion of the Scotch pine in Northern Europe.

Hustich, I. (1966). *Ann. Univ. Turku* A II **36**, 7–47. On the forest-tundra and the northern tree-lines.

Ingstad, H. (1966). "Land under the pole star". St Martin's, New York.

Ingstad, H. (1969). "Westward to Vinland". Macmillan, Toronto.

Jungerius, P. D. (1969). *Arctic Alpine Res.* **1**, 235–245. Soil evidence of postglacial tree line fluctuations in the Cypress Hills area, Alberta, Canada.

Kleiselsierg, R. von (1947). *Ber. naturw-med. Ver. Innsbruck* **47**, 9–32. Die heutige Schneegrenze in den Ostalpen.

Klikoff, L. G. (1965). *Ecol. Monogr.* **35**, 187–211. Microenvironmental influence on vegetation pattern near timberline in the central Sierra Nevada.

Koch, L. (1945). *Meddr Grønland* **130**, 1–374. The East Greenland ice.

Kozubov, G. M. and Shaydurov, V. S. (1965). *Akademiia Nauk SSSR Izvestiia*, Ser. Geog. **3**, 101–104. Vertical zonality in the Khibin mountains and fluctuations of the timberline (in Russian).

Kriuchkov, V. V. (1958). Moscow Universitet. Geograficheskiĭ Fakul' tet Informatsionnyĭ Sbornik o Rabotakh po MGG No. 3, 29–40. Boundaries of woody vegetation as indicators of climatic conditions (in Russian).

Kukla, J. (1969). *Geologie Mijnb.* **48**, 307–334. The cause of the Holocene climate change.

Laitakari, E. (1920). *Acta for. fenn.* 17(1) 53 pp. Untersuchungen über die Einwirkung der Witterungsverhältnisse auf das Längen- und Dickenwachstum der Kiefer (in Finnish).

LaMarche, V. C. and Mooney, H. A. (1967). *Nature* **213**, 980–982. Altithermal timberline advance in western United States.

Lamb, H. H. (1959). *Weather* **14**, 299–318. Our changing climate, past and present.

Lamb, H. H. (1963). UNESCO Arid Zone Res. 20, 125–150. On the nature of certain climatic epochs which differed from the modern (1900–39) normal.

Lamb, H. H. (1964). *Q. J. R. met. Soc.* **90**, 382–394. Trees and climatic history in Scotland.

Lamb. H. H. (1965). *Palaeogeogr., Palaeoclimatol., Palaeoecol.* **1**, 13–37. The early medieval warm epoch and its sequel.

Lamb, H. H. (1966). "The Changing Climate". Methuen, London.

Lamb, H. H. (1967). *Geogrl J.* **133**, 445–468. Britain's changing climate.

Lamb, H. H. (1969a). *Nature* **223**, 1209–1219. The new look of climatology.

Lamb, H. H. (1969b). *Revue Géogr. phys. Géol. dyn.* **11**, 363–380. Activité volcanique et climat.

Lamb, H. H., Lewis, R. P. W. and Woodroffe, A. (1966). R. Meteorol. Soc. Internat. Symposium (World Climate from 8000–0 B.C.). Imperial College, London, Proc. pp. 174–217.

Lawrence, D. B. (1950). *Geogrl Rev.* **40**, 191–223. Glacier fluctuation for six centuries in southeastern Alaska and its relation to solar activity.

Löve, D. (1959). *Can. J. Bot.* **37**, 547–585. The postglacial development of the flora of Manitoba: a discussion.

Manley, G. (1961). Intern. Assoc. Sci. Hydrology, Publ. No. 54, 388–391. Meteorological factors in the great glacier advance (1690–1720).

Marr, J. W. (1948). *Ecol. Monogr.* **18**, 117–144. Ecology of the forest-tundra ecotone on the East Coast of Hudson Bay.

Maycock, P. F. and Matthews, B. (1966). *Arctic* **19**, 114–144. An arctic forest in the tundra of northern Ungava, Quebec.

McKelvey, P. J. (1953). *N.Z. Jl For.* **6**, 435–448. Forest colonisation after recent volcanicity at West Taupo.

Mikola, P. (1952). *Fennia* **75**, 69–76. The effect of recent climatic variations on forest growth in Finland.

Molloy, B. P. J. (1969). *J. Hydrology* (New Zealand) **8**, 56–67. Evidence for postglacial climatic changes in New Zealand.

Molloy, B. P. J., Burrows, C. J., Cox, J. E., Johnston, J. A., Wardle, P. (1963). *N.Z. Jl Bot.* **1**, 68–77. Distribution of subfossil forest remains, eastern South Island, New Zealand.

Moss, E. H. (1932). *J. Ecol.* **20**, 380–415. The vegetation of Alberta IV. The poplar association and related vegetation in central Alberta.

Nicholls, J. L. (1956). *N.Z. Jl For.* **7**, 17–34. The ecology of the indigenous forest of the Taranaki upland.

Nichols, H. (1967). *Eiszeitalter Gegenw.* **18**, 176–197. The post-glacial history of vegetation and climate of Ennadai Lake, Keewatin and Lynn Lake, Manitoba (Canada).

Onthank, D. G. (1951). *Ecology* **32**, 730–731. Forest development following glacier recession observed in 1776.

Ording, A. (1941). *Meddr Skogstorsvoes* **7**, 1–354. Årringanalyser på gran og fura.

Oswalt, W. H. (1960). *Tree-Ring Bull.* **23**, 3–9. The growing season of the Alaskan spruce.

Patten, D. T. (1963). *Ecol. Monogr.* **33**, 375–406. Vegetational pattern in relation to environments in the Madison Range, Montana.

Pears, N. V. (1968). *Oikos* **19**, 71–80. The natural altitudinal limit of forest in the Scottish Grampians.

Porsild, A. E. (1938). *Trans. R. Soc. Can.* **32**, 21–38. Flora of Little Diomede Island in Bering Strait.

Porter, S. C. and Denton, G. H. (1967). *Am. J. Sci.* **265**, 177–210. Chronology of neoglaciation in the North American Cordillera.

Raeside, J. D. (1948). *Trans. R. Soc. N.Z.* **77**, 153–171. Some post-glacial climatic changes in Canterbury and their effect on soil formation.

Rampton, V. (1969). "Pleistocene Geology of the Snag-Klutlan Area Southwestern Yukon, Canada". Ph.D. thesis, Univ. Minnesota.

Raup, H. H. (1937). *J. Arnold Arb.* **18**, 79–117. Recent changes of climate and vegetation in southern New England and adjacent New York.

Raup, H. H. (1941). *Bot. Rev.* **7**, 147–248. Botanical problems in Boreal America.

Raup, H. H. (1966). *Meddr Grønland* **166**, 1–112. The structure and development of turf hummocks in the Mesters Vig district, Northeast Greenland.

Rekstad, J. (1901). Norges geologiske andersøgelses aarboy for 1902 (3), 1–48. Iagttagelser fra braeer i Sogn og Nordfjord.

Richmond, G. L. (1962). U.S. Geol. Survey Prof. Paper 324, 135 pp. Quaternary stratigraphy of the La Sal Mountains, Utah.

Ritchie, J. C. (1960). *Can. J. Bot.* **38**, 185–197. The vegetation of northern Manitoba IV. The Caribou Lake Region.

Samuelsson, G. (1910). *Bull. geol. Inst. Univ. Uppsala* **10**, 197–260. Scottish peat mosses. A contribution to the knowledge of the late-quaternary vegetation and climate of North Western Europe.

Sauberer, F. and Dirmhirn, I. (1950). *Wett. Leben* **2**, 248–261. Die Bedeutung des Strahlungsfaktors für den Gletscherhaushalt.

Schofield, J. C. (1964). *N.Z. J. Geol. Geophys.* **7**, 359–370. Post-glacial sea-levels and isostatic uplift.

Schove, D. J. (1954). *Geogr. Annln* **36**, 40–80. Summer temperature and tree-rings in north Scandinavia A.D. 1461–1950.

Schove, D. J. and Lowther, A. W. (1957). *Medieval Archaeology* **1**, 78–95. Tree-rings and medieval archaeology.

Schuurmans, C. J. E. (1965). *Nature* **205**, 167–168. Influence of solar flare particles on the general circulation of the atmosphere.

Schuurmans, C. J. E. (1969). Konink. Nederlands Meteorol. Inst., Mededel. en Verhand. No. 92. The influence of solar flares on the tropospheric circulation.

Shulman, M. D. and Bryson, R. A. (1965). *J. Appl. Meteorol.* **4**, 107–111. A statistical study of dendroclimatic relationships in south central Wisconsin.

Siegel, S. (1956). "Nonparametric Statistics for the Behavioral Sciences." McGraw Hill, New York.

Siren, G. (1961). *Commun. Inst. For. Fenn.* **54**, 1–66. Pine at the timberline as an indicator of climatic fluctuations in Fennoscandia in historic times (in Swedish).

Slåstad, T. (1953). "Årringundersøkelser i Gudbrandsdalen". Thesis, Univ. Oslo.

Steensberg, A. (1951). *Nature* **168**, 672–674. Archeological dating of the climatic change in North Europe about A.D. 1300.

Strand, J. (1962). *Det. Norske Skogforsøksvesen* **18**, 41–84. Temperaturendringer i de siste decenniene.

Suess, H. E. (1965). *J. Geophys. Res.* **70**, 5937–5952. Secular variations of the cosmic-ray-produced carbon 14 in the atmosphere and their interpretations.

Terasmae, J. and Anderson, T. W. (1970). *Can. J. Earth Sci.* **7**, 406–413. Hypsithermal range extension of white pine (*Pinus strobus* L.) in Quebec, Canada.

Thorarinsson, S. (1956). "The Thousand Years Struggle Against Ice and Fire". Reykjavik: Museum Nat. History Misc. Papers No. 14. 52 pp.

Tikhomirov, B. A. (1962). *Polar Rec.* **11**, 24–30. The treelessness of the tundra.

Tikhomirov, B. A. (1963). *Can. Geogr.* **7**, 55–71. Principal stages of vegetation development in northern U.S.S.R. as related to climatic fluctuations and the activity of man.

Turner, J. (1964). *New Phytol.* **63**, 73–90. The anthropogenic factor in vegetational history, 1. Tregaron and Whixall Mosses.

Tyrrell, J. B. (1896). *Grol. Surv. Can. Ann. Rept.* New Ser. **9** pp. 1–218. Report on the Doobaunt, Kazan and Ferguson Rivers and the North-west coast of Hudson Bay.

Uspenskii, S. M. (1963). *Priroda* **52**, 48–53. Warming up of the Arctic and the fauna of high latitudes (in Russian).

Vebaek, C. L. (1962). In Proc. Conf. Climates 11th and 16th Centuries, Aspen, Colo., June 16–24. Natl. Center Atmospheric Res. Tech. Notes 63, 17. Material from south-western Greenland.

Wardle, P. (1963). *N.Z. Jl Bot.* **1**, 301–315. The regeneration gap of New Zealand gymnosperms.

Wardle, P. (1968). *Ecology* **49**, 483–495. Engelmann spruce (*Picea engelmannii* Engel.) at its upper limits on the Front Range, Colorado.

Wells, P. V. (1970). *Science* **167**, 1574–1582. Postglacial vegetational history of the Great Plains.

Willett, H. C. (1950). Centenary Proc. R. Meteorol. Soc. pp. 195–206. Temperature trends of the past century.

Willett, H. C. (1951). *J. Meteorol.* **8**, 1–6. Extrapolation of sunspot-climate relationships.

Willett, H. C. (1961). *Ann. N.Y. Acad. Sci.* **95**, 89–106. Patterns of solar climatic relationships.

Yamamoto, T. (1961). *J. met. Soc. Japan* Ser 2, **39**, 269–281. Sunspot–climatic relationships in fluctuations of glaciers in the Alps and atmospheric precipitation in Korea.

Author Index

Numbers in italics refer to the pages on which references are listed in bibliographies at the end of each article.

Subject Index

D

T

Y

Z